EUROPE/AMERICA
6

A HIGH TECHNOLOGY GAP?

EUROPE, AMERICA AND JAPAN

why is a US group worried about Europe's loss of competitiveness? "such a Europe would be unconfident, pro-protectionist, politically unstable & Eastward-looking" p35

Good to see CFR's project on European-American relations recognizing centrality of Japan to this process. Unfortunately, only the Japanese contributor notes rapid emergence of E Asian NICs as serious forces in world economy.

Best analysis of Europe's problems & solutions comes from de Benedetti — not just size/industry split & market fragmentation by nat'l bodies, but also Eurosclerosis: rigidities by law & social custom that prevent responding quickly enough in rapidly changing industries.

EUROPE/AMERICA
6

A HIGH TECHNOLOGY GAP?
EUROPE, AMERICA AND JAPAN

Andrew J. Pierre, Editor
Frank Press
Hubert Curien
Carlo De Benedetti
Keichi Oshima
Introduction by Robert D. Hormats

A Council on Foreign Relations Book

Published by
New York University Press
New York and London
1987

COUNCIL ON FOREIGN RELATIONS BOOKS

The Council on Foreign Relations, Inc., is a nonprofit and nonpartisan organization devoted to promoting improved understanding of international affairs through the free exchange of ideas. The Council does not take any position on questions of foreign policy and has no affiliation with, and receives no funding from, the United States government.

From time to time, books and monographs written by members of the Council's research staff or visiting fellows, or commissioned by the Council, or written by an independent author with critical review contributed by a Council study or working group are published with the designation "Council on Foreign Relations Book." Any book or monograph bearing that designation is, in the judgment of the Committee on Studies of the Council's board of directors, a responsible treatment of a significant international topic worthy of presentation to the public. All statements of fact and expressions of opinion contained in Council books are, however, the sole responsibility of the author.

Printed in the United States of America.

**First published in hard cover by New York University Press,
Washington Square, New York, N.Y. 10003**

Library of Congress Cataloging-in-Publication Data
A High technology gap?

(Europe/America ; 6)
1. Technological innovations—Economic aspects—Europe. 2. Technology Transfer—Economic aspects—Europe. 3. Technological innovations—Economic aspects—United States. 4. Technology transfer—Economic aspects—United States. 5. Technological innovations—Economic aspects—Japan. 6. Technology transfer—Economic aspects—Japan. I. Press, Frank. II. Series.
HC240.9.T4H54 1987 338'.06 87-6713
ISBN 0-8147-6600-5

The Project on European-American Relations

Relations between Western Europe and the United States have become more turbulent in recent years. Divergences in interests and perceptions have grown. Many are questioning the fundamental assumptions of the postwar period. There is a broad consensus that the European-American relationship is in a state of transition.

A new generation is emerging and a number of social and cultural changes are under way that are also contributing to this transition. While our common heritage and values set limits on how far we may drift apart, there is an increasing recognition of the divergences between the United States and Europe on such critical issues as defense and arms control, policy toward the Soviet Union, East-West trade and technology transfer, West-West economic relations, North-South issues, and problems outside the NATO area. The challenge for statesmen will be to manage the differences—and where possible create a new Western consensus—in such a way as to enable the Alliance to adapt to new circumstances while preserving its basic character.

The relatively simple world of the postwar period is gone. Americans today appear to have less understanding of European perspectives and Europeans less appreciation of American views. There is much handwringing about the trans-Atlantic malaise, but less constructive thinking about how to manage and, where possible, reduce our differences.

The project is designed to identify and clarify the differences in interests and perspectives affecting critical issues in the European-American relationship, thereby enhancing understanding across the Atlantic. Approximately three issues per year are selected for examination on a rolling basis over a three-year period. The issues are those that are most likely to create friction in the period ahead.

A short book is published on each issue. European and American authors with points of view that differ from each other but represent important strands of thought in their respective societies contribute analyses of the problem and offer their policy prescriptions. We hope that by disaggregating the issues in this manner, we can make a constructive contribution to the Atlantic debate.

An advisory group of Council members, with the participation of European guests, helps to choose the issues and discusses the ideas in the manuscripts prior to publication. They are, however, in no way responsible for the conclusions, which are solely those of the authors.

We would like to thank the Rockefeller Foundation, the Andrew W. Mellon Foundation, the German Marshall Fund of the United States, and the Ford Foundation for their assistance in supporting this project.

Cyrus R. Vance

Advisory Group
Project on European-American Relations

Cyrus R. Vance, *Chairman*
Robert D. Hormats, *Vice-Chairman*
Andrew J. Pierre, *Director of Project*
Kay King, *Assistant Director of Project*

David L. Aaron
George W. Ball
Seweryn Bialer
John Brademas
Hodding Carter, III
Robert F. Ellsworth
Murray H. Finley
Richard N. Gardner
Stanley Hoffmann
Robert E. Hunter
Irving Kristol
Jan M. Lodal
Charles S. Maier
Robert S. McNamara
Harald B. Malmgren
Maynard Parker
William R. Pearce

Robert V. Roosa
Nathaniel Samuels
J. Robert Schaetzel
John W. Seigle
Marshall D. Shulman
Robert B. Silvers
Anthony M. Solomon
Helmut Sonnenfeldt
Joan E. Spero
Ronald Steel
Fritz Stern
John R. Stevenson
John H. Watts, III

Paul H. Kreisberg, *ex officio*
Peter Tarnoff, *ex officio*

The editor would like to thank C. Michael Aho, Georges P. Berthoin, Peter Cowhey, Wolfgang Danspeckgruber, John Diebold, William Diebold, Ellen Frost, David Gompert, Charles Herzfeld, Jerome Jacobson, Joshua Lederberg, Rodney Nichols, Victor Rabinowitch, Charles Schmitz, Pierre Simon, Eugene Skolnikoff, Mike Yoshitsu and John Zysman for their assistance in planning and/or commenting on the manuscripts. He would also like to thank Moira Coughlin, Meg Hardon, David Kellogg, and Rob Valkenier for their assistance in the production of this book.

The Project on European-American Relations is under the auspices of the Council's Studies Program.

Already published:

Nuclear Weapons in Europe, edited by Andrew J. Pierre, with contributions by William G. Hyland, Lawrence D. Freedman, Paul C. Warnke and Karsten D. Voigt.

Unemployment and Growth in the Western Economies, edited by Andrew J. Pierre, with contributions by Marina v.N. Whitman, Raymond Barre, James Tobin and Shirley Williams, and an introduction by Robert D. Hormats.

Third World Instability: Central America as a European-American Issue, edited by Andrew J. Pierre, with contributions by Fernando Morán, Irving Kristol, Michael D. Barnes, Alois Mertes and Daniel Oduber.

A Widening Atlantic? Domestic Change and Foreign Policy, edited by Andrew J. Pierre, with contributions by Ralf Dahrendorf and Theodore C. Sorensen.

The Conventional Defense of Europe: New Technologies and New Strategies, edited by Andrew J. Pierre, with contributions by Andrew J. Pierre, Richard D. DeLauer, François L. Heisbourg, Andreas von Bülow and General Sir Hugh Beach.

Contents

About the Authors

Frank Press is President of the National Academy of Sciences. He served as science adviser to President Jimmy Carter and Director of the Office of Science and Technology Policy from 1977 to 1981. A member of the faculty of MIT from 1965 to 1977, he also taught at the California Institute of Technology from 1955 to 1965. Mr. Press has participated in bilateral science agreement negotiations with China and the Soviet Union and was a member of the U.S. delegation to the nuclear test ban negotiations in Geneva and Moscow. He is the co-author of the textbook, *Earth*.

Hubert Curien served as France's Minister of Research and Technology from 1984 to 1986 and is currently a professor on the Faculté des Sciences at the University of Paris. He is also Chairman of the Scientific Council for Defense. Prior to his tenure in the Cabinet, Mr. Curien was President of the Centre National Etudes Spatiales from 1976 to 1984, General Delegate for Scientific and Technological Research from 1973 to 1976, and Director-General of the Centre National de la Recherche Scientifique from 1969 to 1973. Formerly President of the European Science Foundation from 1979 to 1984, he also served as Chairman of the Council of the European Space Agency from 1981 to 1984.

Carlo De Benedetti is Chairman and Chief Executive Officer of Olivetti & Co. He is also Vice Chairman and CEO of Compagnia Industriali Riunite (CIR) and the Italian financial group Compagnia Finanziaria De Benedetti (COFIDE), as well as Chairman of Buitoni, the Euromobiliare Finance Company and Sasib. Mr. De Benedetti serves as Vice Chairman of the General Confederation of Italian Industry and is a member of the Roundtable of European Industrialists. Previously, he was Chairman and CEO of Gilardini and CEO of Fiat. From 1983 to 1986 he acted as Chairman of the Council for the United States and Italy, where he is still a member of the Executive Committee.

Keichi Oshima is Chairman of Technova, Inc. in Japan; Vice Chairman of the Industrial Research Institute; and Professor Emeritus at the University of Tokyo, where he taught from 1950 to 1981. In addition, Mr. Oshima sits on the Advisory Committee on Industrial Technology to Japan's Ministry of International Trade and Industry (MITI) as well as on the Prime Minister's Advisory Committee on Overseas Economic Cooperation. He served as a member of the Advisory Committee of the U.N. Center for Science and Technology Development from 1981 to 1986 and as Director for Science, Technology and Industry at the Organization for Economic Cooperation and Development from 1974 to 1976. His publications include *The American and Japanese Auto Industries in Transition.*

Robert D. Hormats is Vice President for International Corporate Finance of Goldman, Sachs & Co. and a Director of Goldman Sachs International. Previously, he served as Assistant Secretary of State for Economic and Business Affairs from 1981 to 1982; as Deputy U.S. Trade Representative (with ambassadorial rank) from 1979 to 1981; and as Deputy Assistant Secretary of State for Economic and Business Affairs from 1977 to 1979. Mr. Hormats was a Senior Staff Member for International Economic Affairs on the National Security Council from 1974 to 1977, having been on the staff earlier from 1969 to 1973. In addition, he was a member of the U.S. delegation to the Versailles, Ottawa and Venice Economic Summits and played a major role in earlier economic summits.

Andrew J. Pierre is a Senior Fellow at the Council on Foreign Relations and the Director of the Project on European-American Relations. Formerly on the staff of the Brookings Institution and the Hudson Institute, he has taught at Columbia University. In addition, he served with the Department of State as a Foreign Service Officer in Washington and abroad. Mr. Pierre is the author of *The Global Politics of Arms Sales, Nuclear Politics: The British Experience with an Independent Strategic Force, Nuclear Proliferation: A Strategy for Control,* and other works.

A HIGH TECHNOLOGY GAP?
EUROPE, AMERICA AND JAPAN

Robert D. Hormats

Introduction

We are living in the midst of a scientific and technological revolution that has transformed virtually every aspect of our existence. It has brought about dramatic advances in medicine, agriculture, manufacturing, military weaponry, and the dissemination and use of information.

Such advances have been made possible largely by the development of the computer, which has permitted the collection, analysis and utilization of information on a scale and with a precision undreamt of only a few years ago. But technological success today also depends on a host of other factors such as a strong educational infrastructure, a large number of highly trained scientists and engineers, a financial system that provides funds and rewards for successful entrepreneurs, laws and economic policies that support the process of innovation, the willingness of labor and management to innovate, close ties between the scientific and industrial communities, and social attitudes that encourage the taking of risks. Because technological and scientific success in a nation results from a combination of forces, it cannot be attained merely by the production or import of computers or other pieces of modern equipment. If that were so, disparities among nations could be reduced relatively easily. Rather, progress is both built into and the product of a nation's attitudes and institutions.

Three major characteristics of the high technology revolution make it strikingly different from the considerable industrial, scientific and technological changes that took place earlier in this century.

First, it is highly dependent on advances in knowledge and information as opposed to heavy fixed investment and large inputs of labor on which much of the economic growth of earlier years was based. Today a premium is placed on education, research, the adaptability of labor and management, creativeness in designing and commercializing new products, and the skill to utilize new technologies to lower the cost of and improve old products. The cost of making a semiconductor, for instance, is largely accounted for by expenditures on research, testing, quality control, and product application—only 10 to 15 percent is labor. And the advanced machinery involved is of little use without highly trained personnel to run it. In many sectors companies are moving rapidly to replace manual workers with very sophisticated machines that require considerable knowledge and training to run; intensify the use of high value-added personnel; invest in research, information, communications and entrepreneurial skills; and, reward innovations in labor and management with financial incentives.

Second, spectacular progress is being made, not just in a few areas but on many fronts at the same time, because various breakthroughs are interacting with and reinforcing one another. The computer cuts the time necessary to undertake complicated experiments in medicine and physics. Highly precise and sensitive instruments enhance research and development in new technologies like fiber optics and synthetic alloys. Communication breakthroughs permit instantaneous exchange of information among scientists and researchers as well as between them and industrial users of their products. Basic research is increasingly being integrated with applied research. Teams of doctors, scientists and engineers combine their disciplines to collaborate on new products such as the artificial heart, while physicists and engineers together explore the magic of the laser in manufacturing and in space.

Third, progress has an important international component. Earlier industrialization was based on "national champions" in the automobile, steel and other industries. International competition was of a milder form, with shifts in competitiveness occurring relatively slowly. Today competition in the race to dominate the new technological frontiers, or to commercialize products that others have developed first, is intense; and borders mean less. Technology and information are constantly being transmitted across national boundaries, and simultaneously companies in many nations are making enormous efforts to be the first to use them profitably and build market share before others catch up. But while competition is strong among nations in the general sense that technological breakthroughs of industries located within a nation's borders create a number of important economic and other benefits, a new type of competition is also emerging. This involves, not national industries but international corporate alliances, created to spread the cost of research, rationalize production, or tap a large number of markets quickly in order to reap the rewards of success. And even where formal alliances do not exist, companies based in any one nation produce components and final goods for companies in a number of other countries depending on such considerations as exchange rates, wages, and proximity.

Single-nation industries are becoming a thing of the past, even in the United States with its large domestic market. And protected industries, which have little incentive to develop trans-border linkages or to bolster their world competitiveness, are likely to fall behind their foreign competitors. Sheltering high technology companies imposes an especially high cost for an economy because it deprives domestic purchasers of state-of-the art technologies from abroad. This in particular hurts traditional sectors that depend on new technologies to modernize.

The importance of new industrial and technological advances now, as in past history, extends well beyond their immediate benefits—as impressive as these are. Johannes Gutenberg's typographical revolution in the mid-fifteenth century made books available to millions of people who had previously found them prohibitively expensive. By broadening access to the written word it made possible far greater popular exposure to new ideas, many of which challenged commonly held assumptions and doctrine. It thus contributed mightily to pressures that, over time, dramatically transformed both church and state. Society and politics were never the same again.

The industrial revolution that began in England in the eighteenth century had an equally profound effect on civilization. It revolutionized not only the production process but also the way man organized society and how he fought wars. It changed forever the economic and political face of Europe, led to rapid urbanization in many parts of the world, contributed to the growth of the United States, and altered the character of life throughout the planet.

Why This Study?

Like the economic and technological revolutions of the past, the remarkable discoveries of scientists, engineers and medical researchers in recent years—and the formidable advances they have made in biological, informational, and material technology—are having a profound impact on society. Recent discoveries are leading to the deindustrialization of the labor force as high technology enables the same unit of output to be produced by fewer workers. New technology is permitting the dispersion of the production process because it encourages the creation of small enterprises and facilitates communication among factories, offices and homes, thus reversing the concentration of the production process of past years. It will also alter the relative economic strengths of nations, their ability to defend themselves, and the overall quality of life of their citizens.

Those nations that failed to participate in the industrial revolution fell behind those that did in virtually all facets of their existence—and some over time ceased to exist as nations at all. In the technological race of today, falling behind also imposes heavy economic, social and security costs. It is, therefore, understandable that modern nations want to excel in that race.

Leadership in the development and application of new technologies must clearly be a priority for the United States; and by many measurements the United States has done well. The vigorous recommitment of its industries to research and development, increasingly innovative management, closer ties between universities and business, the cold reality of foreign competition, and a surge in entrepreneurialism have produced strong results in key sectors.

But it is also important that our close friends and allies achieve advances in and take advantage of these technologies. We in the United States might, on the one hand, resent the aggressive competition of other Western nations and want to stay ahead of them in research and the commercialization of new technologies while protecting ourselves against predatory trade practices. On the other hand, we would suffer if they were to fall so far behind as to cause a significant weakening of their security, their economies, or their social fabric. Our foreign markets, our alliances, and our own well-being would ultimately be weakened.

This is why we have undertaken a study of Europe's technological competitiveness. Concerns have been expressed on both sides of the Atlantic that Europe is suffering from a "technology gap" vis-à-vis the United States and Japan in some important sectors. To this it should be added that a number of Americans perceive a serious and growing gap between the United States and Japan in some industries—but that is not the main focus of this particular study.

In this volume we ask four basic questions:

- Is there a technology gap between Europe on the one hand and the United States and Japan on the other?
- What are the reasons for the gap?
- What are the implications for Europe and the United States of a large and continuing technology gap?
- How can the gap be narrowed?

Is There a Technology Gap?

The conclusion reached by all contributors to this volume was that despite considerable European success in important sectors such as nuclear energy, aerospace, biotechnology, computers, and some other significant areas of information technology, a technological gap does exist between Western Europe and its major industrialized-country trading partners.

Frank Press candidly put the point as follows: "If one accepts the proposition that future economic viability will depend on *the ability to introduce a broad array of advanced technologies across all economic sectors in a cohesive region of critical size,* then the aggregate list of separate European national strengths is not impressive and the state of affairs is worrisome. From this point of view a technological gap exists."

Carlo De Benedetti notes that "this gap—measured in terms of higher unemployment rates, higher costs, lower production capacity, and lower structural competitiveness in high-tech sectors—has created great concern about the future of Europe's industry and economy."

What Are the Reasons for the Gap?

There was also a broad consensus on why this gap exists. It is, De Benedetti suggests, "more and more widely understood that Europe's weaknesses are a result of excessive market fragmentation, national protectionism, inefficient and costly public intervention, and a reduced capacity to compete in the new international cycle." For example, although European R&D spending is "almost double that of Japan and about two-thirds that of the United States," in many cases "R&D initiatives are duplicated, so that resources are unproductively utilized, or research is mainly carried out by public laboratories and universities without any link-up with industry." De Benedetti also points out that "national champion policies in high-tech sectors have failed to promote a competitive European industry, because their fundamental premise is incorrect. . . . The protected company feels less need to achieve greater competitiveness by investing in new technologies, developing new products, entering new markets, or reaching new international alliances."

Concurring with this assessment, Hubert Curien stresses that "the lack of a truly integrated European market" is a serious drawback for European industry. "The disparity in habits and in the size of each piece of [the] European puzzle demands many energy-consuming efforts from which Americans have been spared by both history and geography."

Frank Press underscores his concerns about the lack of a "trans-European response to competition with the United States and Japan." Europe, he fears, "is not realizing a full return on its investment in science." Europe's problem, he asserts, "is the absence of political leadership that would expedite a long overdue techno-economic integration and would address the sociotechnical factors . . . that impede innovation. If these circumstances persist, the technological gap may never be closed, resulting in significant economic consequences."

Curien cites another impediment to European competitiveness—the lack of interest displayed by most European banks in ventures in the high technology realm. "Banks merely project their clients' existing interests and tendencies. European citizens do not really like to venture into industry. They know little of their own industries and do not instinctively feel proud of them." He adds that "the middle-class morals of Europeans lead them to consider failure as not just an accident but as a sin, and even worse, a sin that no confession can easily absolve. While it is often argued that the laws of European countries do not sufficiently favor the development of venture capital, a closer examination shows that legal and fiscal arrangements that would encourage this kind of investment are already numerous. The real problem is that eagerness, more than capacity, is lacking."

What Are the Implications of the Gap?

The consequences of a failure to narrow the technology gap are wide-ranging and severe. Keichi Oshima notes that "high-technology industries have become a crucial ingredient in a nation's performance in international trade and economic activities." A technology gap, he states, "is recognized as having a direct impact on social stability and military affairs, thus threatening West European security vis-à-vis the Eastern bloc."

One observer, cited by Press, remarked that a Europe which was technologically and economically noncompetitive would be "'unconfident, protectionist, politically unstable, and Eastward-looking.'"

European industries lagging in the technology race would continue to suffer from, in De Benedetti's words, an "inability to participate on an equal basis in international bargains"—which would put them at a critical disadvantage in many fields from computer technology to strategic defense.

How Can the Gap Be Narrowed?

Both Europe and its major industrial partners need to make a greater effort to reduce the technology gap. As a founding father of the "Eureka" program, Curien feels it represents an important step toward closing the gap. "Those of us who helped to found Eureka believed that further stimulation of European industry through an essentially market-oriented move would be welcome. Industrialists are better and more precisely acquainted than politicians and administrators with the realities of the market's requirements. They have the means to assess in real terms the evolution of market demands and their clients' expectations." Curien stresses that "the initial idea was to build a program where decisions would be made essentially from the bottom up—that is, with governments taking their cues from industrial and commercial agents."

De Benedetti feels that "an entrepreneurial breeze is blowing through Europe emanating from southern countries, like Italy, but pervading the whole continent. Investments in manufacturing automation, R&D, and technological innovation have made a strong recovery. European cooperation in R&D and production has been boostged by the introduction of the ESPRIT [information technology] and RACE [telecommunications] programs."

To further the process De Benedetti advocates pursuit of four objectives. First, he proposes "the integration of European systems at all levels (including monetary and economic policies)." Europe, he states, must be a "unified global competitor" for "without a truly common market open to competition and without a reasonable level of integration among financial, fiscal, judicial and social structures, the European nations are bound to lose ground technologically and strategically." Second, he advocates the development of "a unified capital market, with a common European currency, Eurostocks, and common tools to channel savings toward innovative companies." Third, De Benedetti urges "the promotion of a decisive entrepreneurial approach, not just in companies, but in government and public service as well." The final objective he outlines is "greater scientific development and utilization of technology based on closer cooperation among companies, universities, and public laboratories and on technological partnerships with U.S. and Japanese companies."

Press argues along similar lines that "the future economic viability of Western Europe depends on a much higher level of political and economic integration. . . . By aggregating European

R&D capacity, capital resources, markets, government procurements, and human resources, the scale of Europe's response would be significant." He also points out that "innovation not only requires new knowledge but the willingness and ability of management and labor to use that knowledge." And political leadership will be required "to persuade the citizenry to accept the attitudinal changes needed for improved labor productivity and the acceptance of new technology."

Press suggests ways in which the United States can help. "Opening SDI and other defense-related contracts to procurement in Europe" was "a step in the right direction." However, he adds, "restrictive U.S. policies for the export of the most advanced technology on national security grounds have grated on allies as have past attempts to cancel contracts with European subsidiaries and partners over policy differences on the export of technology to non-NATO countries and the Soviet bloc."

Japan also has a role to play. "Its contribution to the world bank of scientific knowledge is disappointing," says Press. "In a real sense Japan has withdrawn more than it has contributed. Japan is now increasing its basic scientific research budgets as a competitive measure. It has an obligation to contribute even more across all scientific fields rather than emphasizing only those with commercial potential." Press also makes a number of suggestions as to how Japan might address complaints about the openness of its markets. He enjoins all nations to work together to remove trade barriers and work toward an improved system of patents and copyrights.

Oshima points out that "Japan has been entangled in the complex web of a technological alliance with the United States, predominantly in the civilian area but recently extending to military technology as well." Thus, "Euro-Japanese cooperation in technology among both private firms and governments is extremely weak compared to U.S.-Japanese cooperation." He asserts that "the Japanese government insists that Japan's catch-up era is now over; that it should contribute to the accumulation of international common scientific and technological knowledge; and that it has to reverse its role from a technological borrower to a giver by transferring Japanese technology to other countries—to Western Europe and even to the United States."

Conclusion

Although this study has attempted to answer questions we believed to be central to a deeper understanding of Europe's technology gap, the ultimate answer to the problem lies in the hands of Europeans themselves. And throughout our deliberations we shared a confidence that with sound policies, creative approaches, and a European-wide effort the gap can be narrowed and ultimately eliminated.

In recent years, too much emphasis has been placed on Europessimism. Little of that exists in fact. However, there has been a tendency in many quarters in Europe to believe that economies would grow and create jobs no matter how high the taxes, how pervasive the disincentives, and how complacent the attitudes of workers and managers toward efficiency and international competitiveness. Cold economic realities have led to deep changes and greater realism. Policies are becoming more market oriented, internal European barriers are being torn down, and workers' wage demands have moderated. Moreover, a surge of entrepreneurialism has emerged in Europe, much as it has in the United States, and Europe's great corporations are again assuming an international leadership role and moving toward the vanguard in many of the new technologies.

Europe has many strengths—human, cultural and economic—that can produce vigorous technological and economic progress. Indeed, the underlying message of this study is that Europe's technology gap, while a problem, need not be an enduring one if complacency, intra-European restrictions, and fragmented R&D programs can be overcome; if private sectors in Europe are given latitude to work with one another and with companies in the United States and Japan with a minimum of government encumbrance; and if Europe's economic partners do their share to strengthen two-way flows of research and information.

The cost of a failure to close the technology gap are high, while the benefits of a technological surge by Europe are enormous—making achievement of that goal of urgent importance, not only for Europe, but also for the United States and Europe's other friends who share an interest in the prosperity and security of that continent.

January 1987

Frank Press

Technological Competition and the Western Alliance

The remarkable advances in science and technology now underway are having profound impacts on the industrial, agricultural and service sectors of the world economy, impacts that will be even greater in the years ahead. Some have called this the "golden age" of science and technology and have termed its impact as a "historic economic transition."

Within that generalization, it is difficult to make specific predictions. However, on the levels of firms, nations and regions there are both good and bad performers in the initial stages of this transition. And the rapid changes characteristic of the technological revolution means that new entrants can displace current leaders. Further, although most advanced nations are committed to a future based on technological progress, few political leaders understand the process of innovation. Without a vision of the technological future and a strategy for achieving a secure position, untutored policymakers can place their nations at economic and political risk. Indeed failure to achieve technological strength and economic success by any of its partners can place the Western Alliance at risk.

The economic impact of advanced technology is all the greater because its benefits will not flow solely from the production and marketing of "high-tech" products. Instead the greatest benefits may flow from the diffusion of advanced technology throughout the industrial, agricultural and service sectors. In industry, the potential of advanced technology to optimize design and manufacturing, reduce retooling and inventory costs, and increase productivity and product quality opens the possibility of rejuvenating sunset industries and restoring many of the jobs lost to foreign competitors. In agriculture, costs can be lowered by introducing genetically engineered crops that are resistant to pests and droughts and that require reduced amounts of fertilizer. The service sectors from health to banking to communications to transportation are ideal candidates for remarkable advances with the introduction of new technology. Indeed the new products evolving from frontier science and technology may be a principal route for the advanced nations to maintain their high wage levels and standard of living. No wonder advanced technology is an agenda item of the economic summits, the subject of multiple initiatives of the European Economic Community (EEC) and individual nations, and a highlight of the platforms of political parties! The public is now aware that how a nation performs in the acquisition and use of science and advanced technology will determine its future economic viability.[1]

In this paper I will explore how Western Europe, Japan and the United States are faring in the new age of science and technology. I will begin by characterizing the rapid progress in science and technology and by describing the nature of the innovation cycle. A key point will be that successful technological innovation depends not only on the excellence of a nation's science and technology base, but also on supportive macroeconomic and political policies, on the quality of management decisions at the level of the firm, and on receptiveness by labor. I will then take up the question of a technology gap between Western Europe on the one hand, and the United States and Japan on the other. Does it exist? If so, why, and what are its consequences? If there are problems, what are the solutions? In a final section I will discuss some policy implications for the United States, Japan and Western Europe. A topic not treated, but as important as any considered, is the multiple roles of de-

---

[1]See, for example, *Global Competition, the New Reality*. Report of the President's Commission on Industrial Competitiveness, Washington, D.C.: U.S. Government Printing Office, 1985; *The Race for the New Frontier: International Competition in Advanced Technology*, New York: Simon and Schuster, 1984; and *The Vision of MITI Policies in the 1980s*, Tokyo: Ministry of International Trade and Industry, 1980.

veloping countries as markets for the new products of the advanced nations, as competitors advantaged by low-cost labor, and as claimants for access to advanced technology.

I. The Golden Age of Science and Technology

Each generation of scientists claims that theirs is a golden age. What makes the present time unique, however, is the rapidity of advance in almost all scientific fields. This unprecedented progress rests on factors specific to our times: the large number of scientists at work, the introduction of instruments of extraordinary sensitivity, the ability to perform experiments on a scale and with a precision heretofore considered unachievable, the ready availability of computers that permit the rapid analysis of large amounts of data, and all of these leading to brilliant syntheses and theories that change the course of entire scientific fields. New fields are being created by combining old ones—for example, molecular biology or surface physics. Interdisciplinary teams of physicists, chemists and engineers create new materials, while physicians, physicists and engineers invent biological imagers to replace exploratory surgery. As traditional fields are being redefined, the boundaries between basic and applied science, and between science and engineering, are gradually eroding. The breakthrough that leads to the next generation of integrated circuits could come from a university physicist or an engineer in industry.

Progress in technology is keeping pace with advances in scientific knowledge. Information transfer between scientist and technologist is being accelerated by new relationships in the United States between research universities and industry. Indeed, the flow of knowledge and know-how is in both directions. In the core technologies of information processing, materials and biotechnology, the time between a scientific discovery and its commercialization can now often be measured in years rather than decades.

It is a characteristic of advanced technology that any listing of subjects changes rapidly. A current list would include:

- electronics—for example, the ubiquity of chips in everything from sewing machines to carburetors;
- computers and microprocessors;
- communications and information processing;
- specialty materials;
- artificial intelligence and robotics;
- airframes and avionics;
- space technology;
- smart weapons;
- computer-aided design and manufacturing;
- biotechnology, including its application to industrial processes, pharmaceuticals, treatment of disease, energy sources, and agriculture;
- catalysis and other chemical processes.

Many of these fields can generate new high-technology products or can help to revitalize low-technology industries.

The effect of technological change on industrial performance is certainly not a new phenomenon. In the United States, economists agree that a major factor in increased productivity is technological innovation—a factor considered even more important than capital and labor quality (see Chart I). That reality was true in the past and will be more so in the future because the technological menu will be all the richer.

II. The Innovation Process

Technological innovation is the process of using new scientific knowledge or better engineering to design, manufacture and successfully market new or improved products. Nations that invest heavily in research and development (R&D) tend to have an advantage in the introduction of new products. However, to the

CHART I

The Role of Technological Innovation in Increasing Productivity

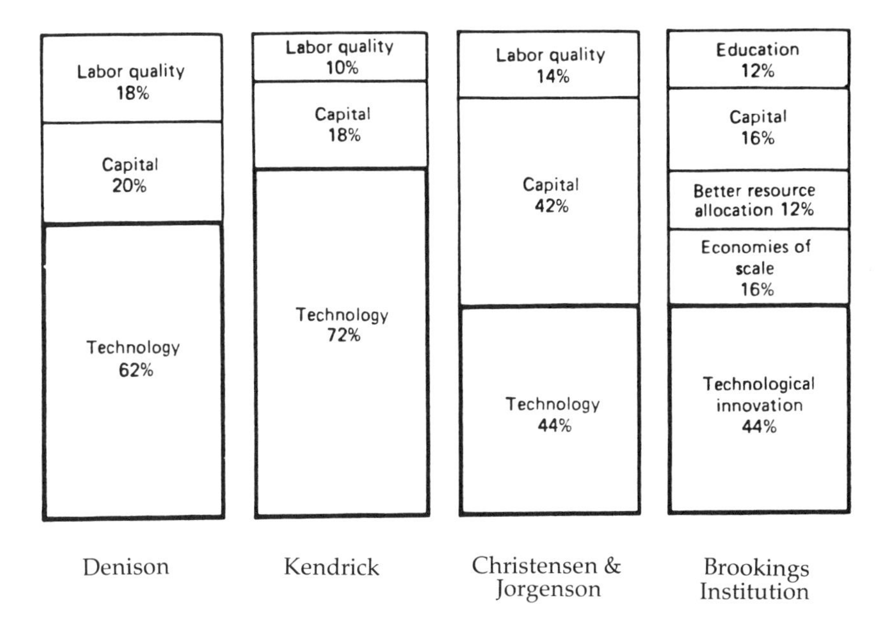

Source: After Bruce Merrifield, as reported by the U.S. Department of Labor in "Productivity and the Economy," *Bulletin of the Bureau of Labor Statistics,* no. 1926, 1977, p. 63; and in *Spectrum,* October 1978, p. 46.

consternation of many inventors, such products as fiber optics, carbon-fiber reinforced plastics, monoclonal antibodies, integrated circuits, jet engines, magnetic recording, and office copying machines all have one thing in common—they were conceived in one country, yet other countries lead in their commercialization.

The lesson of that experience is that successful innovation involves more than scientific creativity and technological ingenuity. For example, economists, innovators and entrepreneurs argue that a nation's macroeconomic policy and other factors, such as currency ratios, interest rates, trade policy, tax policy, antitrust policy, and patent policy are among the determinants of the successful exploitation of advances in science and technology. Obviously, management wisdom and skill at the level of the firm are controlling factors. Less tangible but significant is whether a nation's "culture" encourages or inhibits innovation. Are there entrenched attitudes in the educational system, the government bureaucracy, management, and the labor movement tending to inhibit the introduction of new technologies?

Stated succinctly, a nation's innovative capacity can be gauged by the answers to several questions: Does the environment spur the creation and use of new knowledge and new or improved ways of doing things? Are effective macroeconomic policies in place that provide a healthy climate for investment, growth and commercial success?

I have prepared the matrix shown in Chart II as a convenience in assessing national policies and culture as they might influence innovative capacity. The rows of the matrix list the usual elements of the classical, linear model of the innovation cycle: research, development, production and distribution. The columns show some of the important national policies and attitudes that can affect performance for each of these elements: government science and technology policy, management skills, national macroeconomic policy, and sociotechnical factors. In the last category, borrowing a term used by Harvey Brooks,[2] I include such elements as research infrastructure, attitudes of labor and the bureaucracy, and the national disposition toward saving, risk taking, and social programs. The intersections of the matrix form a convenient outline for a discussion and comparison of national policies and attitudes and their influence on innovation.

The matrix obviously oversimplifies a complex process. For example, a sector-by-sector analysis for industry is preferable and would show major differences. Also, innovation is not a continuous process that flows linearly from research through

---

[2]See "Technology as a Factor in U.S. Competitiveness," in B.R. Scott and G.C. Lodge, eds., *U.S. Competitiveness in the World Economy*, Boston: Harvard Business School Press, 1985.

CHART II
The National Innovation Matrix

A Framework for Comparing National Policies and Capacities for Technological Innovation

	Science & Tech. Policy	Macroeconomic Policy	Management Skills	Sociotechnical Factors
Research				
Development				
Production				
Distribution				

marketing.[3] It can occur at any entry point of the innovation process. As the Japanese have shown in many cases of commercial success, a solution to a difficult development problem rather than a research breakthrough, or improvements in the manufacturing process to increase productivity or quality, can enhance the competitiveness of a product. Often, an improved product is more successful than an innovative one. Salesmen may be the first to perceive the need for a new product, which then triggers new research and developmental efforts. New knowledge generated in one country can lead to successful technological advances in another. In many key fields today, a research break-

[3]The linear model of innovation is an obvious distortion of a complex process but is often used for lack of an improved model. See Stephen J. Kline and Nathan Rosenberg, "An Overview of Innovation," in Ralph Landau and Nathan Rosenberg, eds., *The Positive Sum Strategy: Harnessing Technology for Economic Growth*, Washington, D.C.: National Academy Press, 1986; and Ralph E. Gomory, "Research in Industry," *Proceedings of the American Philosophical Society*, vol. 129, 1985, pp. 26–29.

through leads rapidly to commercial success, often bypassing or moving very quickly through the steps in the innovative cycle. Similarly the entries representing national policy and infrastructure are somewhat arbitrary and certainly incomplete. Again, the point is that the matrix serves only as a framework for a discussion.

Research

The governments of the advanced industrialized democracies have assumed the role of patrons of basic research. The training of scientists and the support of research is justified as a social contribution whose benefits are widely disseminated to the benefit of the nation and its citizens. The level of support of science may vary over time—for example, in recent years it has been significantly up in France and the United States and surprisingly down in the United Kingdom—but the appropriateness of government support is no longer debated.

The private sector generally supports a level of in-house research commensurate with the technological level of the product; those firms that do it well manage research programs so as to lead to timely results but leave long-term projects with uncertain recoverability of costs to universities or government laboratories.

National infrastructure is an important factor in a nation's research performance. A large country has an inherent advantage over a small one. Other positive factors that enhance a nation's scientific performance include: mobility of scientists, that is, easy movement between research centers and between universities and industry; mechanisms for the recognition and rapid promotion of the most creative individuals; access to modern equipment and facilities; ease of scientific communication.

In the United States a unique and advantageous system has evolved with some 2,000 schools offering degrees in science and engineering, of which some 100 are considered research universities by the scale of their research effort and by their successful integration of teaching with research. These schools enjoy good connections to the industrial and agricultural sectors, and to government laboratories.

Overall, by reasons of historic development and public support the large industrial democracies, with differences among them, lead the world in science and technology. There are some indications that on grounds of either protectionism or national security some of these nations may move to restrict scientific communication in key fields such as microelectronics, computers and biotechnology. This could be divisive within the Western Alliance.

Development

In the United States there is a general consensus, somewhat loose in its application, that development is best left to the private sector. It is held that market-driven development tends to be more successful than technology-driven development, and that the private sector is better than government at judging what can be manufactured and sold. Costs can be recovered from the revenue stream of successful products, obviating the need for a direct government role except to provide for such matters as the protection of intellectual property, safety, standards, and antitrust regulations. The latter, for example, prevent American firms from cooperating in technological development clearly identified with commercial products. On the other hand, such indirect government policies as R&D tax credits, low capital gains taxes, accelerated depreciation, and investment tax credits have been used to motivate technological development by the private sector.

The American consensus also allows important exceptions that justify direct and large-scale government involvement in cases where national security and high risk, together with significant social return or other national needs, are involved. Agriculture, energy and defense are the principal examples. It has been noted, especially by foreign competitors, that the exceptions have in the past permitted government underwriting of American leadership in a number of key technologies: microelectronics, computers, communications, space systems, nuclear reactors, biotechnology, farm products, airframes and avionics.

In contrast to the previous decade, the first half of the 1980s have seen in the United States a resurgence of interest by private sector management in R&D investments, as well as a massive increase in defense R&D. Overall, the United States devotes a higher proportion of its gross national product to R&D than does any other country, although West Germany and Japan have higher percentages devoted to civilian R&D. Pointing to the extremely specialized nature of modern military acquisitions, some observers question whether the United States currently gains as much advantage from technology "spillover" from the defense to the civilian sector as in the past. Nevertheless, it is an impressive statistic that in 1983 American industry (including government-supplied funds) outspent Japan and the EEC countries combined for R&D. As to R&D infrastructure, the research university system has provided the United States with the highest proportion of R&D scientists and engineers in the Western world. Further, the easy availability of venture capital and the large equity market in the United States have facilitated the launching of start-up firms that emphasize R&D and the commercialization of new technologies.

Without going into details, many European countries and Japan tend to be more interventionist than the United States about development targeted for the civilian sector. Most of the mechanisms described in the previous paragraph are variously employed by those countries, but so are such mechanisms as government coordination, direct subsidies, relaxed antitrust regulations, and, in some cases, protection against imported technology. The latter may take the form of standard-setting and procurement that favors domestic producers (not that the United States has not flirted with some of these approaches on occasion). Government laboratories, especially in Japan, contribute significantly to technology development identifiable with potential products. Japan's Ministry for International Trade and Industry is a primary source of the sciences and technologies needed for high-technology development. In a sense, MITI is Japan's answer to the U.S. Department of Defense in its contribution to technological development. The institutional structures for conducting R&D in many European countries and Japan may compensate for the fact that their universities are culturally less attuned to industry than U.S. universities.

Although the level of funding is not yet very large ($350 million in 1984), the EEC has undertaken a number of initiatives designed to promote R&D within its member-states. The goals are to develop new techniques and improved products for mature industries and to promote new industries based on advanced technology. A common EEC strategy for R&D is envisioned that would involve share-cost projects carried out jointly by mixed groups of partners, which may be companies, universities or other entities. The initiatives include ESPRIT (European Strategic Program for R&D in Information Technology), BRITE (Basic Research in Industrial Technologies for Europe), and RACE (R&D in Advanced Communications-technologies for Europe).

In a new effort to aggregate Europe's technological capacity, France has proposed a pan-European initiative called "Eureka." Its purpose is to promote cooperation between European research centers and companies in such fields as artificial intelligence, supercomputers and very fast microelectronic circuitry. It is not surprising that the Eureka technologies overlap those of the new, huge U.S. Strategic Defense Initiative (SDI). The motivation may be both exasperation with the slow pace of the EEC response to the technological challenges from abroad and fear of the technological advantage that may be gained by the United States as a result of SDI. As of late 1986 some 19 European countries, including all EEC nations have endorsed Eureka. These efforts are too new to judge whether Europe is on a path toward achieving a common policy and a trans-European dimension to R&D that will replace the separate and subcritical national programs.[4]

Production

In most sectors manufacturing is the costliest component of the innovation process. Without the availability of capital at reason-

---

[4]For a summary of national science and technology policy and indicators of R&D performance see *Science Indicators 1984*, Washington D.C.: National Science Board, 1985; and *Science and Technology Policy Outlook*, Paris: Organization for Economic Cooperation and Development, 1985.

able cost, together with incentives for investment in plants and equipment, technological advances cannot contribute to productivity gains and improved products. In fact, there is little enthusiasm to undertake R&D in the first place.

Whether it be industrial or agricultural products or services, improved ones or new ones, the contest goes to those who have learned how to deliver at low comparative cost and high quality. In view of the substantial rate of increase of real wages over the past decade, which outstripped productivity increases, Western Europe must have a much higher level of productivity improvement to stay in competition. In his article in this volume, Carlo De Benedetti discusses many sociotechnical factors within Europe that will make this higher level difficult to achieve: a shortage in some countries of skillful, entrepreneurial managers knowledgeable about the process of technological innovation; structural rigidity and resistance to change by many labor unions; a cultural aversion to risk-taking because of the high consequences of failure; limited financial tools to raise capital; and a political imperative for governments to protect large companies despite their inability to compete.

A combination of macroeconomic and sociotechnical factors and a unique industrial strategy have worked much to the benefit of Japanese products in world markets.[5] Japan has demonstrated the value of an educated and dedicated work force and skilled management that work together constructively and can adapt easily to new technology. Japanese business also benefits from a culture that values savings that, together with supportive government policies, leads to the availability of large amounts of low-cost capital for investment. Martin Feldstein proposes that the fundamental reason for Japan's trade surplus with the rest of the world is its high savings rate![6] In addition, there is a general perception that in the area of consensus-build-

---

[5]See, for example, Daniel I. Okimoto, "The Japanese Challenge in High Technology," in Landau and Rosenberg, eds., *op. cit.*

[6]Martin Feldstein, "American Economic Policy and the World Economy," *Foreign Affairs*, Summer 1985, pp. 995–1008.

ing, coordination and mutual support between government and business, and business and banking, Japan performs exceedingly well, creating a stimulating environment for economic growth.

Starting with "smokestack" industry products and following with high-technology products, Japan has emphasized up to now the later stages of the innovation cycle, i.e., the lower half of the national innovation matrix (see Chart II). Highest priority was given to putting in place the most advanced process and production technology, aggressive pricing, and mass marketing. Japan was content to be the second to the market with new products, believing—rightly so—that better quality and lower costs of its products would displace those of the original innovators. It confirmed with a vengeance the comment that "pioneers are those with arrows in their backs." Both Daniel Okimoto and I, among others, question whether this unique formula for commercial success will work as well in the future, suggesting that Japan may have to compete head-on with the United States across the entire innovation cycle.[7] As we shall see in the next section, the trend in Japan is to improve performance in research as well as development (the upper half of the innovation matrix) at the same time that the United States is moving to do better in production and distribution.

Despite a few exceptions, such as the controversial but nevertheless self-liquidating and successful protection of the Chrysler Corporation, the American consensus of minimal government intrusion in the production system via direct financial subsidies, "picking winners," or aiding sunset industries in distress, continues. The U.S. government has limited ability to control directly such specific private civil sector outcomes, nor does it want to. Nevertheless, a number of effective, indirect methods of support have been invoked from time to time, including favorable tax policies, government procurement, automobile import quotas, and preferential relaxation of regulations—many of which have stimulated increased private investment. At different times American experience with high interest rates, infla-

---

[7]See Okimoto, *op. cit.*; and Frank Press, *Newsweek*, January 16, 1984, p. 11.

tion, unnecessarily expensive and confusing regulations, and tax policies that were instituted seemingly without regard to their effect on investment, demonstrated how a variety of government policies can weaken competitive positions and deter innovation. As is the case with most countries, agriculture is a special case, with a full array of direct and indirect support mechanisms that have evolved over many decades.

A significant trend is worth noting. In contrast to the previous decade, there are indications that a renewal is under way in the United States to improve manufacturing productivity. Investment by U.S. industry in modern production facilities based on advanced technology has grown rapidly in recent years. This is exemplified by the durable goods industries, especially those in the so-called rust belt or smokestack belt of the Midwest.

This region is experiencing a new momentum and could in time regain many of its earlier losses. The evidence is in the form of large R&D spending and large investments in technologically advanced manufacturing facilities in recent years. Multibillion-dollar investments by the automobile industry and other companies with long traditions of manufacturing in the region are pacing the gradual recovery. Some 16,000 companies now make high-technology equipment in the 8 states bordering the Great Lakes, including robotics, optics, biomedicine, computer software and electronics. This combination of high technology and traditional manufacturing, new for this region, is being encouraged by state governments serving in a new stimulative role and using a variety of incentives, such as access to universities and, in some cases, state funds. These trends augur a regaining of competitiveness in world markets. It remains to be seen whether this will be at the expense of employment levels and the high standard of living of workers in the region.

For political or cultural reasons governments of most other advanced nations are somewhat less timid about becoming involved in direct support of civil sector industry, especially under political pressure of growing unemployment in noncompeting industries. Such interventions can take the form of government investment in or acquisition of firms, subsidies, domestic procurement, selective protection against imports, and export aids. France and Japan, among others, believe that there is no clear break between development and production, and that a government's role should not stop with the support of development but should assure the production of the new product.

Distribution

Exports have been a major factor in the economies of Japan and Europe for a multiplicity of reasons: the need to balance large imports of fuel and other resources; the lack of large domestic markets; attractive product lines with technology superior to that of most of the world. The imperative to export was not as significant to the United States, a nation that has a large, captive domestic market, a wealth of resources, and a technological self-sufficiency.

It was inevitable that foreign firms would see the opportunities of competing with American companies on U.S. turf. Although the lesson was a painful one, American managers no longer view domestic markets as a safe preserve and are now following in the footsteps of their European and Japanese counterparts in seeking global markets, as well as competing with more vigor domestically.

A major source of friction among the advanced nations is conflicting political and economic policies that affect trade, either by design or unintended consequence. The web of national policies and their effects on trade is a complex one, difficult to disentangle. The global consequence of the huge American budget deficit provides a good example: the deficit resulted in high interest rates and high values of the dollar, neither of which was conducive to American innovation and exports. As a result, many European products became attractive to American consumers, and European exports increased. On the other hand, the capital inflow to the United States, attracted by high interest rates, may have deprived Europe of the investments needed for more productive plants and equipment.

Another illustration is the extent to which the United States discourages certain exports of technology and products on political or national security grounds. Examples range from the embargo of agricultural exports to the Soviet Union following the invasion of Afghanistan to difficult licensing procedures for supercomputer sales to some Western governments out of concern of possible access by Soviet-bloc nationals. Most affected firms view these regulations as impediments to their ability to compete.

There is general agreement that the global benefits of free trade outweigh its costs. However, given the pressure to sustain domestic industries and given the variety of mechanisms available to governments and firms to enhance the competitiveness of their national products, free trade has become an elusive goal theoretically worth pursuing, yet not always practiced.

Although one nation's policies under the guise of free trade often appear to be unfair trading practices to another, some general concepts of fairness in trading practices have evolved. Almost all trade negotiators in developed countries would agree that below-cost dumping of products in another country to weaken or destroy competition is unfair, as is government procurement (except for defense industries) restricted to domestic producers. Government export subsidies are in the unfair category, but concessional export credits are in a gray area. Tariffs to protect domestic producers carry the risk of retaliatory actions, but hidden barriers, equally effective, are difficult to prove in the negotiation process and even more difficult to eliminate. There are sociotechnical factors that confer trading advantages to certain nations that are basically not unfair; for example, more productive labor, lower social and defense costs, and high rates of savings. A dilemma faced by political leaders is whether or not to protect an industrial or agricultural sector rendered uncompetitive by national policies or culture rather than by unfair trade practices of other nations. Some nations, including the United States have "escape clause" or "supreme national interest" options even if dumping and predation are not factors.

Trade practices are of special concern for advanced technology, which is increasingly viewed as important to a nation's future economic security if not its national security. The high costs of developing new products, the brief time before copies appear, and rapid obsolescence make for a short innovation cycle—often three to five years. Fast access to global markets is mandatory for generating the income stream needed to sustain the momentum of innovation. For this reason, it is particularly damaging when a nation, with a significant domestic market and the capacity to compete in advanced technology, impedes access to innovative foreign products, for example, by "infant-industry" protection measures, until a domestic manufacturer achieves a competitive position. Henry Rosovsky has called this policy "denial of the profits of innovation." He argues that it has been practiced by Japan to the detriment of American innovation, and that maximum pressure for its removal is warranted.[8]

III. The Technology Gap

There is an abundance of statistics and indicators that in their totality should provide all the information needed to describe and compare the performance of individual firms, business sectors, and national economies. Unfortunately, a theoretical basis for such evaluations is lacking and analysts fall back on heuristic methods in their selection and interpretation of data. Often, interest groups will reach conclusions and make recommendations based on data selected to buttress the case for subsidies, deregulation, protection or other special pleadings. Perhaps this explains some of the conflicting conclusions in many recent reports comparing the technological strength and performance of different countries. Nevertheless this book calls for views on whether there exists a technological gap between Western Europe on one hand and the United States and Japan on the other. For the purposes of this volume, the issue is better framed in terms of a gap in the capacity to generate and use advanced technology across all economic sectors as the key to future commercial success.

The United States

The United States and Japan are the world's largest industrial democracies and the most dynamic countries in the commercial application of technology. Yet the decade of the 1970s saw the

---

[8]See Henry Rosovsky's op ed article, "Trade, Japan and the Year 2000," *The New York Times*, September 6, 1985, p. A23.

United States in a downward trend in a number of indicators of comparative national performance: investments in plant and equipment, productivity growth rates, market losses in sectors of earlier preeminence, and the beginnings of a merchandise trade deficit. This decline, primarily in the lower half of the national innovation matrix of Chart II, was variously attributed to government policies, labor-management conflicts, management weaknesses, the shock of the oil-price rise, and unfair trading practices of others.

To the astonishment of the world, the 1980s have seen a resurgence of the U.S. economy, the strongest economic rebound in the postwar years. Some 10 million (net) new jobs were created, with an additional 4 million unfilled jobs currently being advertised. Over 90 percent of these jobs are in new small businesses, 600,000 of which are being created each year. Although only about 10 percent of the new jobs are in the high-technology category, each position sustains a support structure of several conventional workers. The United States has been viewed as the locomotive of recovery for its allies!

In recent years U.S. manufacturers, in response to their poor competitive performance in the 1970s, and stimulated by favorable tax policies, have increased their R&D expenditures and have embarked on investments in productivity-enhancing equipment at a pace not seen since the 1950s. The American economy now leads in the consumption of high-technology equipment as a percentage of the total of U.S.-Japan-EEC consumption. State-of-the-art numerical machine tools, computer-automated production lines, robotics, and computer control of production flow are beginning to transform American manufacturing. Firms are being restructured to enhance productivity. These changes are occurring across many industrial sectors, involving both traditional manufacturers and the new, advanced technology industries. Although measures of productivity are controversial and uncertain, one indicator—improvement in worker output per hour—is attributed by the Department of Commerce to these investments and to heightened management attention to manufacturing process and production technology and corporate restructuring. The United States is on a path of improving its deficiencies in production, the weakest element of its national innovative capacity in the previous decade.[9]

All of this happened despite the historically high value of the dollar in the period from 1980 to 1985 that contributed to an unprecedented deterioration of the U.S. trade balance in manufactures. Most observers believe that lower interest rates and the declining value of the dollar in time will lessen the impact of what was one of the most serious threats to America's economic future.

Overall, we see the American resurgence driven by the traditional strengths of the United States in the upper half of the innovation matrix coupling with growing strength in the lower half. The following list summarizes these strengths:

- world leadership in science and technology;
- strongest infrastructure for R&D (number of scientists and engineers, level of budgets, number of research universities, coupling between universities and industry, spillover from defense and space technology);
- largest domestic market;
- largest capacity to generate investment capital;[10]
- a mobile and well-trained labor force, with declining unit labor costs in recent years;
- a management selection system based on performance;
- a resilient, large economy that has created millions of new jobs, primarily in the service sector, yet virtually maintained job levels in manufacturing.

The United States is not without its weaknesses and has stumbled on frequent occasions. The huge budget deficit and nega-

---

[9]See L.H.Olmer, *U.S. Manufacturing at a Crossroads,* U.S. Department of Commerce, International Trade Administration, Washington, D.C., 1985; and *Handbook of Labor Statistics,* Table 91, Department of Labor, Bureau of Labor Statistics, Washington, D.C., June 1985.

[10]Of particular interest is the generation of capital for new companies that amounted to $28 billion in venture capital and $25 billion in new public stock offerings by start-up companies since 1981.

tive trade balance have to be dealt with. However, in view of the evidence of a basically resilient economy, steady improvements in the efficiency of production, and a continuing position in the forefront of science and technology, the United States has to be considered the number one seed in the future competition for economic leadership based on advanced technology.

Japan

The United States and Japan have pursued different paths to leadership in the uses of advanced technology. Japan's remarkable success, at first with traditional manufacturing and later in the advanced technology sectors, emphasized the application of good engineering to both manufacturing processes and production technology, and the release of high-quality, competitively priced products on world markets. Although Japan initially acquired many of the necessary technologies from abroad, it knew how to use them, and the improvements it made upon them were significant. Government policies, management skills, advantageous sociotechnical factors, and aggressive trading practices combined to make Japan the best performer in the bottom half of the national innovation matrix. The following list notes some of the Japanese strengths that have been variously cited:

- political stability;
- access to low-cost capital;
- dedicated, skilled, lower-cost work force;
- highly competent management;
- best example of cooperation between government, banks, business and labor;
- large domestic market;
- relatively low social and defense costs.

The Japanese are the first to recognize that past accomplishment is no guarantee of future success. Newly industrialized countries, "mini-Japans" such as South Korea, are making inroads into the traditional manufacturing sectors by a combination of good technology and even lower-cost labor. (The last quarter of 1986 saw Japanese steel companies in a loss position.) The Japanese see the United States moving to correct its weakness in production efficiency and offering greater competition, at least in U.S. markets, if not globally. Thus, Japan sees its future shifting away from mature industries to the commercial rewards that will flow from forefront scientific fields and advanced technology.

Japanese performance in science and in the creation of new technologies has not been as impressive as one might expect from a technological giant. This is a potential weakness in view of the short obsolescence times for the advanced technology products of the future and the decreasing intervals between scientific discoveries and commercial application. However, Japan is instituting programs to correct past shortcomings in science and technology, i.e., to improve its performance in the upper half of the national innovation matrix.[11] Okimoto summarizes a number of actions underway in Japan to prepare for its "high-high tech" future.[12] Prime Minister Yosuhiro Nakasone is personally involved with steering educational reforms and curriculum changes through the bureaucracy. Basic research budgets are being increased, and expanded university-industry cooperation is being encouraged. Many attributes of the free market-approach to innovation, those supposedly responsible for American scientific-technological breakthroughs and new product concepts, are beginning to appear in Japan. These include venture capital markets, greater labor mobility, changes in industrial policy in the direction of less government intervention, and small ventures as subsidiaries of large corporations. Under international pressure, Japan is deregulating its highly structured financial system. Freeing up market forces in this way may well be a boon to creativity and innovation in Japan.

Thus, the 1980s are seeing changes in the United States and Japan in which each country is moving to correct weaknesses il-

---

[11]See MITI, *op. cit.*

[12]Okimoto, *loc. cit.*

luminated by the corresponding strengths of the other. It is as if the separate paths toward technological leadership that have evolved in each country are beginning to converge. This is not surprising as each country positions itself in a confrontation for techno-economic success based on forefront science and the most advanced technology.

Japan's record of success may also pose problems in that its huge trade surpluses are increasingly viewed in the United States and Europe as distorting world markets. If Japan chooses to be responsive, it will redound to reduced competitiveness of many of its products in the years ahead. Nevertheless, Japan's potential for achieving its espoused goal of a future based on the most advanced technology must be considered very high.

Western Europe

Although European science is strong, communication between the research community and the commercial sector is poor. In this sense Europe is not realizing a full return on its investment in science. A trans-European response to competition with the United States and Japan in advanced technology is neither in place nor an agreed-upon goal. At the present time the EEC level of funding of R&D does not exceed some 15 or 20 percent of the total of separate national efforts, and one would expect that there is little transfer of technological know-how from one country to another.[13] The whole of European technology is less than the sum of its parts, many of which are individually impressive.

In the absence of a technologically unified Europe, one can only compile a list of European strengths by looking at the sum of separate national efforts: pharmaceuticals and chemicals in West Germany; robotics in Great Britain; nuclear power, commercial rockets and perhaps aircraft in France; telecommunications, and some other fields in other countries. Such fields as

---

[13]Françoise La Fontaine, "The Government Role in Research and Development: What Other Nations Are Not Doing. The European Community Case." Remarks prepared for a Workshop at the National Academy of Sciences, November 21–22, 1985.

computers and information processing, biotechnology, microelectronics, electro-optics, materials, and office equipment are examples of key technologies not yet on the list. Individual countries—West Germany for example—have proven records of technological strength and economic success. However, many of the mature industries, the traditional core of earlier European economic strength, will face growing, powerful competition from the newly industrialized countries. If one accepts the proposition that future economic viability will depend on *the ability to introduce a broad array of advanced technologies across all economic sectors in a cohesive region of critical size,* then the aggregate list of separate European national strengths is not impressive and the state of affairs is worrisome. From this point of view a technological gap exists and there may be truth in "Europessimism."

Consider the data compiled by John M. Marcum showing the EEC in an inferior position to the United States and Japan in all of the following categories:[14]

- growth in exports of high-technology products—1975–1984;
- fraction of high R&D products in exports of manufactured goods—1984;
- trade balance in high-intensity R&D products—1983;
- balance of payments for patents, inventions, and industrial processes—1983;
- applications of new technology in manufacturing;
- growth in capital investments in new equipment by firms—1982–1984;
- consumption of advanced technologies such as communications, data processing, semiconductors and robots—1984;
- industrial R&D expenditures as a percentage of industrial domestic product—1983;
- growth in unit labor costs in manufacturing—1983–1984;

---

[14]John M. Marcum, "The Technology Gap: Europe at a Crossroads," *Issues in Science and Technology*, Summer 1986, pp. 28–37.

- availability of venture capital (no data for Japan);
- number of scientists and engineers as a fraction of the labor force—1982;
- enrollments in higher education as fraction of total population—1982.

This disappointing list does not augur well for Europe's techno-economic future. The economic present is not doing too well either, as evidenced by the usual indicators. For example, in contrast to the United States and Japan, steadily growing unemployment is one of the major politico-economic problems of Europe. Added to this is the absence of European counterparts to most of the strengths listed for the United States and Japan in previous pages.

A unified Europe could build the capacity to perform well across all elements of the national innovation matrix. Europe's problem is the absence of political leadership that would expedite a long overdue techno-economic integration and would address the sociotechnical factors mentioned earlier that impede innovation. If these circumstances persist, the technological gap may never be closed, resulting in significant economic consequences.

IV. Policy Implications

The tidal changes of scientific advances and the introduction of new technologies are causing the industrialized nations to rethink their economic strategies. Japan is proceeding methodically and consensually in the wise realization that the formulas for its past successes do not necessarily guarantee a bright future. West Germany, France, the United Kingdom, and other European nations are trying to advance both individually and collectively, a dual approach with internal conflicts that are almost always resolved in favor of the nation-state. The United States, troubled over the role of government, nevertheless has the advantage of the best science and technology infrastructure, a large and resilient economic base, and the capacity to identify weaknesses and, in time, correct them.

The rapidity of technological change couples with the fact that purely domestic businesses beyond a certain scale are increasingly anachronistic. A multiplicity of forces, some mentioned in previous pages, have created competitive world markets and a fluid, global economic system. Although markets are now global, and economic interdependence is recognized, U.S. politics and policies remain largely rooted in the nation-state. The United States as a nation-state, and as a leading member of an alliance, is confronted with difficult and conflicting policy issues as it views the strategic consequences of a Western Europe that may in the future be technologically (and presumably, economically) noncompetitive. As an observer has remarked—"such a Europe would be unconfident, protectionist, politically unstable, and Eastward-looking." In an earlier time the Marshall Plan was conceived to rebuild the economic foundation of America's principal allies. Although a technological Marshall Plan is unwarranted and infeasible in today's climate, the United States and its allies are not without options that would allow wide participation in the new technological revolution. The likelihood that the United States and Japan will fare well in the new technological era is high. Therefore it is important to discuss policy issues for the United States, Japan and Europe that relate to improved European performance.

Issues for Europe

Despite Europe's examples of scientific and technological strength and successful, innovating companies, the incorporation of advanced technology across all economic sectors rather than a selected few is the key to trans-European competitiveness. Separate national programs are likely to be subcritical compared to the efforts that can be mounted in the United States and Japan. Those who hold that the future economic viability of Western Europe depends on a much higher level of political and economic integration than has been achieved are correct.[15] By aggregating European R&D capacity, capital resources, mar-

---

[15]See, for example, C. Michael Aho and Jonathan D. Aronson, *Trade Talks, America Better Listen!* New York: Council on Foreign Relations, 1985.

kets, government procurements, and human resources, the scale of Europe's response would be significant. National policies that impede Europe's competitive stance, such as nonuniform standards and regulations, customs barriers, and protection of noncompetitive "national champion" companies, could be eliminated. All of this would put in place transparent borders between the small European nation-states and an unprecedented level of cooperation in the common interest.

Although the postwar years have seen movement toward the political and economic integration of Western Europe, the process has been contentious and the pace slow. National security concerns might have driven integration, but the successful NATO Alliance with its militarily strong American partner served the purpose. Perhaps it will take the "Sputnik" of political instability and economic dislocation, generated by years of poor performance, growing noncompetitiveness, and unemployment, contrasting with the expanding economies of the United States and the nations of East Asia, to overcome the centuries-old traditions of nationalism.

The political and economic integration of Western Europe must of course be a European initiative. Unfortunately, European efforts to date, more than a quarter century after the signing of the Treaty of Rome, are not impressive. The European Parliament is more symbol than substance. At this time the EEC is viewed as overly bureaucratic, with insufficient budget and political clout, an organization where national self-interest too often dominates the common good. The insistence on national exemptions to a unified market are almost always successful, and those exemptions continue as an obstacle to European economic success. For example, it is estimated that customs barriers are costing European firms more than $10 billion a year, and the lack of a European-wide competition for government procurement penalizes European taxpayers some $35 billion each year.

However, efforts are under way under EEC leadership for a coordinated approach to advanced technology. Some of these efforts have been described earlier. It is too early to assess results, but the resources dedicated to these initiatives appear to be too small, as if the individual countries seem to be betting more on their own efforts rather than cooperative efforts. Whether the EEC or deliberately separate efforts such as Eureka are the instruments of choice remains to be seen. In any case, a most important policy recommendation for Europe is to undertake the institutional change that would facilitate a trans-European response to economic competition, a response that would be led by the unified development and use of advanced technology.

The aggregation of European resources is a necessary, but not a sufficient, condition for success in technological innovation and economic competitiveness. Innovation not only requires new knowledge but the willingness and ability of management and labor to use that knowledge. De Benedetti argues in this volume that the introduction of new technology requires new behavioral patterns by European governments, management and labor.

Young, educated, risk-taking managers tend to be the first to employ new technologies. Europe loses to the extent that the corporate culture does not recognize and promote such talent as well as the Americans do. Risk-taking and innovation typically go together; yet the severe penalties of bankruptcy and the social ostracism of business failures is a feature of European culture that instills an aversion to risk-taking. The rigidity of European labor—the lack of mobility and the resistance to new technologies as a threat to jobs and high wage rates—is also a barrier to innovation that must be faced. As mentioned earlier, wage increases at rates that exceed productivity improvements are prescriptions for noncompetitive products and increasing unemployment. A measure of European political leadership in the years ahead may well be the ability to persuade the citizenry to accept the attitudinal changes needed for improved labor productivity and the acceptance of new technology. It is a powerful argument that the new resources created by technological innovation may be the only means to reverse the loss of jobs, to maintain high standards of living, and to provide for the needs of those dislocated in the process of change.

Economists point out two complementary features of the current economic scene: the transfer of technology is the key to economic growth; and there is an acceleration in the transfer of modern technology across borders, in considerable part due to the growing influence of multinational firms.[16] It follows that transborder networking among firms, including R&D partnerships, parent company-subsidiary relationships, mergers, co-production agreements, cross-licensing, second-sourcing, original equipment manufacturers agreements, and other arrangements could play a significant role in the techno-economic performance of Europe. Such relationships already exist and are expanding, not only within Europe but also between American and Japanese firms with European counterparts. Indeed the possibility of a Japan-Europe techno-economic axis based on networking should not be underestimated. Relationships driven by market considerations rather than government push seem to be preferred by successful companies that have the most to offer as network partners. Certainly non-European firms do not underestimate the contributions of European partners to networking in the way of providing technology, market access and capital.

It may well be that if national policies of the kind described in the preceding paragraphs are slow in their implementation, private sector networking will become the principal vehicle for achieving cohesion and critical mass to keep Europe in the race. It may become the preferred mechanism for successful European firms wary of government intervention. In this case governments would be relegated to the role of simply not getting in the way.

Issues for the United States

America's influence in all these matters is captured by the old joke—Question: "Where does the 700-pound gorilla sleep?" Answer: "Anywhere he wants!" As the world's largest economy and the greatest source of advances in science and technology, the United States is automatically a significant factor in the course of economic progress in Europe and Japan.

---

[16]Edwing Mansfield, "Macroeconomics of Technological Innovation," in Landau and Rosenberg, *op. cit.*

The United States has generally supported West European political and economic integration, though not always in agreement with specific policies of the EEC and other entities. Although the United States has limited leverage at the political level in this matter, it can continue to foster integration through instruments that also serve its own interests, such as opening defense procurement to competition from consortia of European firms, arguing in international forums for common standards and free flow of information across borders, cooperating in establishing large international science research facilities, and raising the level of political and economic consultation with the European Community and other integrative entities.

European governments have been critical of the large American budget deficit and its consequences for high interest rates, distorted exchange ratios, and the outflow of capital to the United States. To some extent the American deficit has played a role in the poor record of investment and the unemployment problems of Europe.[17] It must be a source of considerable anguish to Europeans that the inability of Americans to cope with a domestic political stalemate damages European economic growth.

Europeans are also concerned about the implications of SDI, not only for arms control, but also for its supposed impetus to American scientific and technological strength and the resultant widening of the technological gap. It is a step in the right direction that the U.S. Department of Defense is opening SDI and other defense-related contracts to procurement in Europe. In some recent cases a consortium of firms with major European participation won large contracts competitively, to the benefit of both the United States and the Europeans. However, restrictive U.S. policies for the export of the most advanced technology on national security grounds have grated on allies as have past attempts to cancel contracts with European subsidiaries and partners over policy differences on the export of technology to non-NATO countries and the Soviet bloc. These differences go beyond the export of dual military and civilian-use technology

---

[17]Feldstein, *op. cit.*

and include the use of trade as a lever to influence the policy of adversarial governments.

It seems that the United States has too often been oblivious to the effects of its policies on the interests of its best friends. Concern for its own performance and the lack of an American consensus on industrial policy may well submerge consideration of other nations' economic interests. There are real differences of opinion regarding trade with the Soviet Union, Libya and other adversaries, which need to be addressed within the Western Alliance. However, the many instances of failure to consult on American domestic and foreign policy initiatives raises the image of the insensitive 700-pound gorilla. This is not the way to inspire respect for American leadership and to secure the democratic world's most important alliance.

Issues for Japan

The recent techno-economic history of Japan brings forth admiration, envy and fear among her trading partners. "Japan bashing" is a fashion that often submerges underperformance by the accuser and overlooks deserved rewards for a national innovation system that works well. Nevertheless, there are divisive issues between Japan and the other advanced nations that must be addressed.

As mentioned earlier, for a nation of Japan's standing, its contribution to the world bank of scientific knowledge is disappointing. In a real sense Japan has withdrawn more than it has contributed. Japan is now increasing its basic scientific research budgets as a competitive measure. It has an obligation to contribute even more across all scientific fields rather than emphasizing only those with commercial potential. Since a major fraction of Japanese research occurs in industry, this may be difficult to achieve, and the results may not be readily accessible.

Virtually all American and European governments and firms that do business in Japan complain of barriers that put foreign firms at a disadvantage in marketing products and services. Compared to the rest of the industrialized world, Japan's economic nationalism results in the lowest imports of manufactured goods as a fraction of total imports. There are also complaints about government-approved cartels and import protection that facilitate predatory pricing, dumping and other practices that give many Japanese products a greater share of world markets than would occur under conditions of free trade. To the extent that this denies a fair market share and the profits of technological innovation to foreign firms, Japan should review such practices and alter them where there is legitimacy to the complaints.

With respect to Japan's huge trade surplus, fundamentally due to laudable national attributes mentioned earlier such as high savings rates, Japan's obligations are not necessarily remedial of a wrong but are a social contribution to global economic stability. Japan has taken a number of steps in this direction: deregulating its financial markets, constructing manufacturing facilities abroad, opening procurement to foreign firms, stimulating domestic consumption, and exhorting business and the public to purchase foreign products. However, the pace is slow and the results are uneven. Japan must still face the policy issue of how much it is willing to give up in order to respond more positively to world pressure.

Trade Policy: Goals for the Future

Countries that confront each other in a trade war as antagonists make for poor allies. Moreover, a "zero-sum" mentality can lead to destructive trade policies that could deny the world the fruits of advanced technology. Only in recent years have issues of advanced technology appeared in General Agreement on Tariffs and Trade (GATT) discussions or on the agendas of the economic summits, unfortunately without much substance. The issue is a difficult one because nations now see their futures related to advanced technology, and protective instincts are a natural response to those who perceive themselves to be in the threatened or inferior positions. The economic consequences of protectionism, globally and nationally, are well known. In the context of this essay it behooves policymakers to consider the deleterious effect of protectionism on technology transfer through the relatively efficient channels of commercial transactions and networking.

Ideally, one might propose trade guidelines in terms of the "global good," e.g., recognition that technological innovation is a social contribution to the world and that its special needs, such as rapid access to global markets and realization of the profits of innovation, must be accorded. In this vein one might argue that a nation's success that adds to the world's technological store and to aggregate world production should not be penalized in the absence of patently unfair trading practices. However, such a constructive approach to trade policy would be achieved more readily if the negotiators are not excessively disparate in their technological capacity. Although the pace is very slow, Japan and the United States, operating from positions of strength, will most likely move toward these goals in the years ahead and emphasize national innovation policies that enhance performance, as well as networking or alliances between firms of both countries, rather than destructive trading practices.[18]

Current policies in Western Europe are uneven in their treatment of advanced technology imports, with some nations overtly protective and others supporting open-market conditions. A successful trans-European effort for technological innovation, along the lines discussed earlier, is probably the prerequisite to a unified trade policy, one influenced by considerations of the European, if not the "global good."

A constructive goal that members of the Western Alliance might pursue in their own common interest is a more rational, equitable and consistent world system of patents and copyrights. The legitimate protection of intellectual property can both foster technological innovation and facilitate its dissemination and use throughout the world. In the absence of such a world system, reliance on trade secrets, prohibitions on visits,

---

[18]See, for example, *Challenges and Opportunities in United States-Japan Relations: A Report Submitted to the President of the United States and the Prime Minister of Japan by the United States-Japan Advisory Commission, 1984.* Nevertheless the friction between the United States and Japan is currently heated. For example, the U.S. government has taken the unusual step of filing legal action against Japanese firms for allegedly dumping 256,000 random-access memory (RAM) chips at below-cost prices.

exchanges, communications to scientific meetings and publications, and other protectionist actions will grow, denying timely access to new technologies by many potential beneficiaries.

The world trading system is under great pressure, and the possibility of a major breakdown should not be underestimated. The United States, Canada, Japan and the European Communities, recognizing the seriousness of the situation, have called for a new round of trade negotiations under GATT auspices. The issues of concern go well beyond those considered here. However, nowhere is the welfare of the advanced nations so interwoven as in their common dependence on science and technology for the maintenance of their standard of living, let alone their economic progress. It therefore behooves the convenors to recognize the rapidly growing significance of advanced technology in world trade and the special vulnerability of high-technology products to trade practices. If GATT fails to address these issues, another forum will be necessary. Strong efforts should be made to define guidelines appropriate for these products so that the "race for the new frontier" ends, as in the Olympics, with all contenders uplifted.

Hubert Curien

The Revival of Europe

The title of this essay, "The Revival of Europe," is intended to both acknowledge a lag in European technology that should be corrected and to express faith in the deep-rooted strength of the Old World. "Euro-pessimists" are more inclined to analyze Europe's weaknesses with self-indulgence rather than looking for ways to cure them. "Euro-optimists," among whom I count myself, believe the continent has plenty of resilience and try to initiate mechanisms for success in their respective countries.

Competent experts have already diagnosed Europe's shortcomings. A great deal of writing has been devoted to the technology gap in several essential industrial fields between Europe on the one hand, and the United States and Japan on the other. In this essay I will content myself with briefly stating the main characteristics of Europe's problems and then examining to a greater extent the steps that are being taken to cure the diseases with which Europe is inflicted. I hope I will be forgiven for giving particular emphasis to the remedies to which I have personally made a contribution.

The Sources of Europe's Shortcomings

The lack of a truly integrated European market must be *first* on the list of Europe's illnesses. One sometimes uses the term "common market" when discussing the European Economic Community (EEC). A truly common market is still a goal, but one that requires the achievement of matching fiscal, legal, and even social policies. But most European nations, working on the not unnatural assumption that their own systems are better than those of their neighbors, have not yet succeeded in attenuating their differences. Some progress has been made. The Commission of the European Communities in Brussels displays great determination in pursuing this aim. But the disparity in habits and in the size of each piece of this European puzzle demands many energy-consuming efforts from which Americans have been spared by both history and geography.

A *second* reason for Europe's difficulties in the high technology realm is the absence of harmonized markets downstream and the lack of coordination in technological development upstream. The structure of European industry is such that very often several organizations or firms undertake the same kind of development separately. This redundancy is not necessarily an evil in limited or medium-sized enterprises; it can even become a stimulus. But as far as major technological developments are concerned, the lack of coordination means delays and higher prices instead of a better-structured economy. This scattering of efforts is one of Europe's plagues.

A *third* impediment to the expansion of new technologies in Europe is the excessive hierarchy of its industrial organization. The European system of relationships between large and small firms is based much more on the subordination of small firms to large ones than on the combination of all their capacities. Large companies often subcontract to small ones, but, most of the time, the nature of such deals does not allow subcontractors the opportunity to demonstrate their capabilities to the full, especially where innovation is concerned.

The lack of interest displayed by most European banks in ventures in the high technology realm is a *fourth* deterrent. Banks merely project their clients' existing interests and tendencies. European citizens do not really like to venture into industry. They know little of their own industries and do not instinctively feel proud of them. In addition, the middle-class morals of Europeans lead them to consider failure as not just an accident but as a sin, and even worse, a sin that no confession can easily absolve. While it is often argued that the laws of European countries do not sufficiently favor the development of venture capital, a closer examination shows that legal and fiscal arrangements that would encourage this kind of investment are already numerous. The real problem is that eagerness, more than capacity, is lacking.

Moving further upstream, from the financing of industry to the organization of research, Europeans come up against a *fifth* problem, which again stems from the deficiencies of their scattered continent. The sum total of public nonmilitary expenditures devoted to research in Europe is close to that of the United States and higher than that of Japan. But these assets yield less output since their implementation lacks unity and efficiency. Similarly, European scholars are not really organized as a community with easy and natural exchanges. The scholar's Europe is still in search of itself.

In the United States, Defense Department contracts do a great deal to support research and development: two-thirds of public funds for R&D are of military origin. In Europe, military credits amount to less than one-third of total R&D spending. It must be acknowledged that, at least in a few fields of technology, these important injections of military funds have allowed breakthroughs that would otherwise have been much slower, if not impossible. Many major American industrial firms would not have acquired their strength in civilian markets had they not been intensely active in military production.

The *sixth* and final impediment is that fundamental research in Europe does not satisfactorily advance into technical development. In France, for instance, the structure of academic research does not sufficiently encourage this kind of transition. And the quantity of research carried out in an industrial environment, compared to the amount of research undertaken in the public sector, is much lower in Europe than in the United States or Japan. European industry is just not immersed in research.

I have chosen to pass a rather forbidding judgment in this analysis, because I think the quest for solutions must be built on a total lack of self-indulgence. It is also easier to add finesse to a rough sketch than it is to look for defects in an excessively polished analysis. My examination of Europe's shortcomings must take into consideration the fact that all European countries are far from similar and that not all European industrial concerns face the same problems. Europe, fully conscious of its weaknesses but also of its strengths, is, in my view, moving toward a revival.

The Condition of European Technology: The Case of Outer Space

To better understand Europe's technological strengths and weaknesses, it is worth examining the conception and evolution of the programs for exploring and exploiting outer space on each side of the Atlantic. American success in outer space, which must not be belittled by recent misfortunes, is obvious proof of U.S. technical mastery. As early as the 1960s, it became evident that America's prestige, as well as the Soviet Union's, was closely tied to achievements in outer space. The two countries' constant success in outer space projects has been one of the most remarkable technological facts of this century. Confronted with this titanic and two-sided competition, Europe could choose between one of two options: either to become a spectator, who cheers at the more exciting moments, or to become a third player.

For Europe to have chosen the first option would have meant giving up the privilege of being counted among the more technologically advanced nations of the world and accepting the role of second fiddle. Renunciation was inconceivable for all sorts of economic and psychological, as well as civilian and military, reasons. In fact, Europe had no choice but to take an active part in the exploitation of outer space. However, the scope of this attempt was such that it compelled a collective action—no European nation was able to undertake a solo effort in this competition without totally distorting its allocation of resources for research and technical development in its national budget.

As early as 1960, several European countries came together to set up a common program for outer-space activities. But the first years of this effort met with numerous and important problems. The enterprise very nearly came to an abrupt end several times. In retrospect, one can learn a great deal from an examination of the European space program's difficult birth.

One of the essential mistakes was to believe that juxtaposition and integration meant the same thing. Since the French, British and West Germans each had fragments of rockets in 1960, the decision was made to use these fragments to build a whole launcher, in the way that Noah's sons had built the Tower of Babel. The rocket *Europa* was the result of this attempt. Unfortunately, it never could launch any satellites.

This failure was hard to accept but helpful in the long run. Technological Europe was going through its apprenticeship. The setup organized to build *Ariane* was totally different. The development of this rocket was the responsibility of a single taskmaster: the French National Center for Space Studies (CNES). Thus, programming for *Ariane* was developed using standard procedures and led to the production of a commercial product, allowing Europe to win half the world market for launchings.

It may be useful to comment upon this categorization of *Ariane* as a commercial product. The rocket was developed and built under CNES supervision, for the European Space Agency (ESA). This agency was created in 1975 to implement the space programs of 13 European countries and to define a common space policy. It is an emanation of the various member governments; its administrative council consists of civil servants who, in each country, are in charge of space policies. Such an agency, if it is well run (which the ESA is), can be very efficient as far as scientific and development programming is concerned. However, such an organization is ill-equipped at marketing and negotiating the sales of commercial goods. Each man to his own trade. An intergovernmental agency possesses neither the habits nor the facilities of an industrial or commercial firm.

As soon as *Ariane* became operative, it seemed natural to commit its building and sales to another organization. Thus a private company called Arianespace was created to carry out these activities. Although such a decision seemed obvious, its implementation was not so easy. Some national representatives within the ESA council felt frustrated at the prospect of giving up some of the responsibility for *Ariane*, even though they did not have the capacity to assume it with any degree of efficiency. European ambitions have their share of ambiguity; but the political desire for unification, without which nothing can be done, must not lead to uneconomic structures. It is not enough that national administrations are fond of Europe; their fondness should be neither excessive nor exclusive. Thus *Ariane* was steered satisfactorily through a labyrinth of European structures. But there is another aspect to this story that serves to shed light on the relationship between Europe and the United States in the realm of outer space.

After the failure of *Europa,* European countries, apart from France and Belgium, were about to give up on the construction of a space rocket on a commercial scale. All Europe's fears about exclusion from outer space seemed allayed by the notion that Europe could always petition the U.S. National Aeronautics Space Administration (NASA) to launch its satellites. This reasoning was correct but oversimplified, as circumstances were rapidly going to show.

France and West Germany had come to an agreement at the end of the 1960s to build telecommunications satellites together. The satellites were called *Symphonie* and were to be put into orbit by *Europa*. The program for developing these satellites was launched, but at the same time the *Europa* rocket failed. France and West Germany then attempted to sign a contract with NASA to launch two *Symphonie* satellites. The contract negotiations were very illuminating for Europeans. NASA officials naturally behaved as partners who held a world monopoly on commercial launching. It would have shown great naïveté on the part of the European negotiators to assume that NASA would behave differently. Business is business, and friendship does not alter the case. The conditions France and West Germany had to accept in order to obtain these launchings were such that the Europeans could not commercialize the *Symphonie* satellites in a way that would make them competitive with telecommunications satellites held by American interests. In time the American requirements turned out to be beneficial to Europe (and this can be said without irony or bitterness). They allowed Europeans in the space community to persuade still undecided leaders that Europe would have no economic future in outer space as long as it could not enjoy an independent launching capacity. Thus, the commitment to the development of *Ariane* was firmly reinforced.

The same mood prevailed in Rome in 1985 when the Europeans set up a new space program that involved, among other things, the development of a heavy launcher, *Ariane 5*, and a space shuttle, *Hermes*, as well as participation in the building of the American space station. Since the decision to take part in the U.S. space station project was made, discussions between the ESA and NASA to settle the conditions of European participation have been difficult. Why these problems? Because Europeans are not really interested in taking part in the construction of a station in which they would be tenants rather than co-owners, and because Americans hardly want a junior partner who would interfere with decisions for which the United States very naturally wishes to be solely responsible. But it is probably better that these problems come to light sooner rather than later.

Since Europe decided to answer positively to NASA's offer to take part in building a space infrastructure, the issue of a man-carrying vehicle became immediate. Here, too, Europe could have allowed itself to be overly idealistic and become totally dependent on the American shuttle. But how could one conceive of having an apartment in outer space that would be available only with the chief owner's agreement? Participation in the space station project would be meaningless unless Europe could enjoy some autonomy in programming—that is, if it could control its own timetable for access to the space station. Europe, therefore, decided to launch a program for a space shuttle, called *Hermes*, that would be able to dock on the U.S. space station.

This decision was also prompted by a concern for security. It seemed problematic that a large international infrastructure could be serviced by a single means of transportation. This conclusion was reached even before the unfortunate accident that befell the U.S. space shuttle *Challenger* in January 1986. Such an accident and the subsequent interruption in service could happen at a manned orbiting space station. Therefore, a second, independent means of transportation should be welcome.

In the future all means of space transportation, be they American, Soviet, or European, should be developed with the objective of docking on all space infrastructures. Security would be greatly increased, and an atmosphere of international agreement would be encouraged. Talks between Europeans and the

Soviet Union on just such a proposal are under way and seem to be making real progress.

To conclude my discussion on outer space, I would like to point out that scientific, technical, industrial and commercial advancements in Europe must proceed as a "variable geometric" process—i.e., projects should include European countries outside the EEC and at the same time different groups of nations should be encouraged to cooperate on different projects. The building of satellites serves as a good example of this rule. At least half a dozen industrial firms in Europe are capable of building telecommunications satellites. They are far too numerous given the size of the European or even the world market. It is necessary therefore to resort to consortiums, but they must be configured according to the nature of the satellite and the client's identity. In some cases, indeed, it may be more suitable to join European with American ventures than to have a solely European network. An example of this was the satellite *Arabsat,* which was ordered by a group of Arab countries and built by Aerospatiale in France in association with the Ford Motor Company in the United States.

The history of Europe's space program exemplifies the relationship between Europe and the United States in a particularly active technical field—a relationship characterized by mutual trust, but also by a certain degree of realism, which has turned out to be very helpful to both partners. Inasmuch as they are investing about ten times less in outer space than the Americans, Europeans must define their policy in such a way that allows them to adapt to this quantitative discrepancy. Outer space is probably one of the fields where Europe has been walking a tightrope with the greatest success.

The Scientist's Europe

Let us move back from technology to fundamental research. It would be useful to remember that at the end of World War II, "scientific Europe" was exhausted. The exodus of Jewish scholars, who for the most part had fled to the United States, considerably weakened Europe's universities and laboratories. The war left laboratories in a state of great dereliction; only a few areas of research were spared due to their links with war production efforts.

The postwar atmosphere also did not facilitate intra-European scientific relationships. European scholars were naturally attracted to the United States and many went there to complete their studies. Europeans owe their American colleagues a great deal of gratitude for the remarkably generous welcome extended to them. It was chiefly in the United States that scholars from France and Germany, for instance, actually met. For obvious psychological reasons, many Europeans spent several years finding the Rhine more difficult to cross than the Atlantic.

While Europeans rejoiced in the quality of American hospitality, they also acknowledged that it did not particularly encourage them to work toward the unification of the European scientific community. Concern about the lack of a common European scientific community, which did not appear until later, did not emerge from the bottom up (spurred by individual scientists), but filtered from the top down (as the result of economic and political necessities). This may be explained by the fact that scholars are individualists, less sensitive than industrialists to the marketplace or politicians to foreign policy goals.

True, the problem of the disparity of languages and of university degrees has done nothing to simplify the process of building a homogeneous scientific community in Europe. Although scientific degrees can be harmonized by law—and this process is now well under way—the difference of languages is a cultural fact. All attempts to rectify this problem have met with failure and have come up against generally well-grounded criticism. An obvious and, incidentally, enriching solution would be to require European scholars to speak several languages—more than two preferably.

These developments help to explain why European scientists and technicians are slower to become aware of lags in Europe's competitiveness than those who work in the areas of development or production.

What have Europeans been doing in Europe in the last 20 years to build a true community of researchers on the continent? Europe's scientific colleagues in Texas, California or Massachusetts have an obvious advantage. Americans feel part of a strong, highly-reputed and, on the whole, harmonious community. The interchange of ideas within the United States as well as the American employment market for scientists are such that the research community benefits from a stimulation that no single European country could hope to sustain on its own.

Learned societies and professional associations also play an important role for American scientists and engineers, as demonstrated by their high membership figures. The way these associations organize meetings and run scientific publications gives a good measure of their strength. In such a field as physics, for instance, the American Physical Society publishes *The Physical Review*, which draws articles of high quality not only from American authors but from international researchers as well. Such attractions are very difficult to rival; national scientific journals in Europe, with only a few exceptions, cannot compete with their American counterparts, since they do not have a comparable pool of authors to draw upon.

To improve their position, national learned societies in Europe have moved increasingly closer to form a network. Thus, some 15 years ago, the European Physical Society, was created to operate as a holding company for the physical societies of various European countries by overseeing their publications and bestowing the European label on the best of them. In addition, some scientists who live in Europe have contemplated the founding of a European association of scientists that would provide an open forum for European researchers and offer support for promotional campaigns, including lobbying for essential goals.

Over and above improved networking among individual scientists, greater coherence is needed among organizations such as the Centre National de la Recherche Scientifique in France, the Deutsche Forschungsgemeinschaft or the Max Planck Gesellschaft in West Germany, and the Consiglio Nazionale delle Richerche in Italy that have been commissioned by their national governments to finance research. In order to give these organizations the opportunity to compare and match their policies the European Science Foundation was founded in 1970.

The ESF, based in Strasbourg, France, operates as a federation and is run by a very small team (20 or so persons); it neither aims to replace national agencies nor to run them in a hierarchical manner. The ESF provides a forum for the interchange of ideas and tries to initiate greater mobility among researchers within Europe.

In 1984 the ministers responsible for scientific policies in various European countries commissioned the ESF to organize research networks. These are structures with little bureaucracy, designed to reinforce cooperation among European researchers in the fields where such cooperation has become compulsory. A network is composed of nodes and links. The main idea is to build new links between existing nodes and to create new nodes only if necessary. This networking is intended to put not only Europe's facilities but also its brains to better use, to facilitate the sharing of instruments, and to organize common high-level postdoctoral training. The aim, of course, is not only to support the fields in which cooperation already exists but to create support for fields, especially the emerging ones, in which cooperation is poor or even nonexistent. There are a number of fields in which networks have been established or will be shortly including behavioral science, gerontology (an increasingly useful science in medically advanced countries), composite materials, and polar sciences. European scientists hope that thanks to these networks, they will reduce the time needed to reach a critical mass in each of these fields. The ESF has also helped to define instruments of research. It initiated the activities of the European Synchrotron Radiation Facility, which is being set up in Grenoble.

Any discussion of the ESF naturally raises the subject of the scientific plants that maintain such expensive equipment as accelerators, wind tunnels and telescopes. The archetypal plant is the European Organization for Nuclear Research (CERN), which was set up in Geneva in the early 1950s. Its scientific success has been universally acknowledged, and it was recently rewarded by the Nobel prize jury.

But CERN's success must not mask the difficulty of creating and sustaining high-quality organizations at the international level. To coordinate ten or more different national partners within the same venture requires patience and a spirit of mutual understanding, both of which quite naturally tend to deteriorate as soon as technical problems arise. It is much easier, of course, for two or three partners to get together than it is for ten or fifteen. Thus it became very tempting for groups of two or three nations to arrange agreements in order to realize large-scale scientific projects.

The construction of a neutron high flux reactor in Grenoble, France was a successful venture involving three countries—France, West Germany, and the United Kingdom. In view of my support for the principle of variable geometry as applied to Europe's space program and my belief that it is desirable to implement it in such programs as "Eureka," it would be illogical for me to criticize this principle with respect to the construction and exploitation of large scientific plants. But the world of basic research is somewhat peculiar; its attitudes are characterized by greater sentimentality than those of the industrial world. Nonetheless, if it were to turn out that, more or less regularly, the largest European countries did not share the responsibilities of elaborating basic programs with the smallest nations, Europe would not achieve inner harmony. This harmony is well worth the price of accepting a few delays and encumbrances, as long as they are not lethal.

The Role of the EEC

Those EEC officials who favor reinforcement of the scientific and technological bases of European industry have been pushing for changes for some time. They should now see some progress as the European Council has decided to add a new technological dimension to the EEC. European heads-of-state have just added some provisions about research and technological development in the "single act," which now defines the community.

The EEC's interference in the field of research and technology can be justified when one considers that the community is trying to build a broader and more competitive economic entity. The community should favor all research that will lead to the establishment of common industrial norms for the products of several European countries. A great European market cannot come to life without a concerted effort on the part of European governments to match industrial norms.

In order to make two-thirds of their economies competitive and to provide employment for 55 percent of their active populations, the EEC nations are dependent on information technology. On the other hand, not one of these countries has a national market for telecommunications that amounts to more than 6 percent of the world market, while the American market amounts to more than 35 percent and the Japanese market to more than 11 percent of the total world market.

Europeans suffer from a threefold problem in the telecommunications field. First, quantitatively, the community is scattered in 12 national pieces in contrast to the single expanding American market. Second, Europe imports 50 percent of its integrated circuits. Finally, since Europe does not speak with a single voice, it is not listened to carefully in international forums that aim to redefine and regulate telecommunications systems.

These challenges, however, have presented the telecommunications industry with the opportunity to become one of the cornerstones of a technical revival in Europe. Some initiatives are under way. The EEC favors the Open System Interconnection that European industry decided to develop. The Integrated Services Digital Network, which should become progressively operational beginning in 1988, will allow Europe to increase its returns to scale: ISDN is an important step toward improving advanced services in telecommunications. This network should become progressively operational beginning in 1988.

A program called RACE (R&D in Advanced Communications-technology in Europe) is responsible for the studies preliminary to the establishment (hopefully around 1995) of an operational pan-European advanced wideband telecommunications network. Technological developments correlated to this initiative have concentrated on high-performing integrated circuits, such as integrated opto-electronics, broadband switching, passive optical components, image-coding, and the technology of display on a large flat screen. Over 100 European companies are already taking part in this common development program. This particular effort in favor of telecommunications is justified not only by the need to respond to very keen American and Japanese competition; it should also allow Europeans to reinforce links between the community's peripheral and central countries and, on a worldwide scale, to narrow the gap between North and South.

ESPRIT (the European Strategic Program for R&D in Information Technology), another important EEC program, was launched in February 1984 and has three main goals: (1) to promote European industrial cooperation upstream wherever possible; (2) to allow European industrialists to avail themselves of technologies that will be useful to them in reinforcing their positions in the next five to ten years; and (3) to establish a united European platform in preparation for movement toward international norms.

ESPRIT research projects are financed half by EEC funds and half by the participants. The total budget for the first five years is $1.5 billion. Research will be precompetitive—that is, upstream of the production process; it will have a deliberately international approach—that is, each project will draw together firms or research agencies from several European countries. A special effort has been made to involve small and medium-sized enterprises in the project. ESPRIT spans such fields as microelectronics, software technology, the architecture of information processing systems, and factory automation techniques. At work on this program are 500 groups—of which half are industrial firms, a quarter are universities and another quarter are public research institutes.

Any discussion of the EEC's scientific and technical initiatives must mention two important programs, BRITE and Stimulation. BRITE (Basic Research in Industrial Technologies for Europe), is in large part the equivalent of ESPRIT in the technical fields, such as new materials and new production techniques, that are not covered by ESPRIT. The Stimulation program is slightly different in nature. It is more concerned with basic research and aims to stimulate cooperation among European scientists by helping them to realize common research projects. This program has a very selective jury, composed of highly reputed scientists. Laboratories that wish to be granted a project within the Stimulation program compete rigorously with each other for supplementary funds and the honor that selection brings with it.

Such are the EEC's initiatives to combat the difficulties facing Europe. Other longer term programs are also supported by the EEC, particularly studies on the production of energy by nuclear fusion. Its annual budget for research and development amounts to about $1 billion. All European scientists wish it could increase regularly. Such an increase, of course, would be easier to obtain if other problems confronting the EEC, such as agriculture, could be solved expeditiously.

Eureka

The above description of various EEC programs suggests that all necessary structures are already in place and that no new programs need be invented. However, the Europeans have established a new system, Eureka. Why?

The answer is that those of us who helped to found Eureka believed that further stimulation of European industry through an essentially market-oriented move would be welcome. Industrialists are better and more precisely acquainted than politicians and administrators with the realities of the market's requirements. They have the means to assess in real terms the evolution of market demands and their clients' expectations. The initial idea was to build a program where decisions would be made essentially from the bottom up—that is, with governments taking their cues from industrial and commercial agents. The establishment of Eureka was in no way meant to reflect upon the programs run by the EEC, where the bottom-up and top-down styles have been well balanced. Eureka was merely the result of the realization that European integration had not been adequately exploited.

An immediate corollary of this bottom-up process is that it cannot operate except in variable geometric structures. Any systematic regulations that encourage the exploitation of projects within defined geographical limits would automatically bring the process to a halt. However, since this mechanism is designed to build technological Europe, it is necessary to group partners from various countries on the continent in each project, provided that there is reasonably balanced representation in the program as a whole.

At any rate, no collective agency in which each nation would have an absolute veto should intervene at any time during the developmental stages of a given project. Whenever it exists in international agencies, such a veto seldom goes unused. This results, at times, in political interventions that deprive technical innovation of its spontaneous character.

Some might raise objections that an institution such as Eureka, which encourages networks of industrialists to conceive and build new products, runs the risk of suppressing competition and could lead to monopolies. Eureka is not intended in any way to suppress intra-European competition; rather it is meant to bring some order into it, paving the way for healthy and constructive competition, while eliminating catastrophic side effects. It is out of the question for Eureka to steer European industry toward a monopolistic situation involving large-scale consumer goods. Eureka can help to prevent a situation in which industrialists become exhausted through the development of identical goods at a high cost in a scattered manner, resulting in the tardy arrival of overly expensive goods in international markets. By tidying intra-European competition and reducing overlap, the continent will be better-equipped to deal with international competition.

Eureka and SDI. The Eureka project was conceived by French President François Mitterrand not long after Ronald Reagan's March 1983 "Star Wars" speech. Following the Reagan speech, U.S. Secretary of Defense Caspar Weinberger wrote to the allies offering to include them in research programs for the Strategic Defense Initiative. While the French government expressed skepticism at the strategic concept of missile defense, it nevertheless stated that French industrialists had every right to accept SDI contracts. This, however, was based on the sole condition that such commitments (for reasons of secrecy and/or exclusivity) not be in conflict with tasks that they must carry out for the national authorities. At any rate, today, European industrialists, be they French, German, British, Italian, or other seem rather disappointed by the number and the nature of the proposals put forward by the SDI Organization.

Eureka can be viewed with some justification as a reply to SDI, especially in light of the fact that, in introducing Eureka to his European colleagues in April 1985, then-French Minister of Foreign Affairs Roland Dumas wrote that only a technological

Europe "would allow, if need be, cooperation on equal terms with our chief international partners, especially the United States and Japan. A subcontractant Europe, working under licenses, would no longer be Europe."

In some ways the SDI and Eureka initiatives are similar, yet they differ widely in a number of essential aspects. In the first place, Eureka is a *civilian* enterprise while SDI is a *military* one. This is an important difference as far as the funds for each program are concerned. It matters less on the other hand to the industrialists who have been commissioned to run them. Both ventures are attempting to develop highly advanced technologies, and the stumbling blocks, whether civilian or military, are similar in nature. In any event, the impetus given to technological development, be it under Eureka or SDI auspices, will benefit the same industrialists.

Another striking difference between Eureka and SDI can be found in their aims. SDI is attempting to organize a *well-focused* system of military defense; Eureka on the other hand has a less definite goal, for it is trying to give a *general boost* to industries in their civilian-oriented activities. However, the "indefinite" nature of the Eureka program does not imply that its efforts are scattered, with the risk that nobody will benefit. Rather, Eureka, through a wise choice and distribution of initiatives, can introduce modernity to the whole of European industry.

The third, and not least significant, difference is that SDI was initiated by *Americans* and Eureka by *Europeans*. Thus, it is natural that, however strong the links of friendship and the political and military agreements between the two continents, their focus will be different.

Background. The Eureka program was officially created in Paris on July 17, 1985, at a meeting of the foreign and science ministers of 17 states (Austria, Belgium, Denmark, Finland, France, West Germany, Greece, Ireland, Italy, Luxembourg, the Netherlands, Norway, Portugal, Spain, Sweden, Switzerland, and the United Kingdom,). Since then, Turkey and Iceland have joined, raising the total number of partners to 19. (The EEC is also a member.)

At its second meeting, held in Hanover, West Germany in November 1985, the group adopted a seven-page "declaration of principles," which stated:

> The objective of Eureka is to raise, through closer cooperation among enterprises and research institutes in the field of advanced technologies, the productivity and competitiveness of Europe's industries and national economies on the world market, and hence strengthen the basis for lasting prosperity and employment: Eureka will enable Europe to master and exploit the technologies that are important for its future, and to build up its capability in crucial areas.
>
> This will be achieved by encouraging and facilitating increased industrial, technological and scientific cooperation on projects directed at developing products, processes and services having a worldwide market potential and based on advanced technologies.
>
> Eureka projects will serve civilian purposes, and be directed both at private and public-sector markets.

While the Paris meeting was very successful in achieving general agreement on overall objectives, the real test came at the Hanover meeting, where the participants agreed to draw up an initial list of ten projects. This meeting also brought to light a few differences in the interpretations given to the aims of Eureka by each country. For instance, France saw Eureka as definitely and firmly market-oriented, whereas West Germany, while agreeing to this principle, also hoped that Eureka would help solve some public problems (such as pollution and its side effects) common to all European countries. The principle of variable geometry helped the participants to find common ground easily. West Germany proposed ecological programs, such as the analysis of pollutants in the upper atmosphere, to which France initially did not want to be committed. Nevertheless, the French and the West Germans were able to come together in specifically industrial projects.

It may be worthwhile at this juncture to offer an explanation (other than the reference to Archimedes) for the choice of the name Eureka. "Eu" stands for Europe, "Re" for research, and "K" for "koordination." Participants were reluctant to allow the

"A" to stand for "agency" because this might have raised images of yet another international bureaucracy. Therefore, it was suggested that the "A" in Eureka represent "action."

Another important issue that had to be settled was the nature and role of Eureka's secretariat and the size of the bureaucracy it would administer. Two options were open: the first, or more intrusive one, would involve the secretariat in overseeing the permanent technical evaluation teams and the second would have it serve only as a clearing house or information center. Fortunately, the second option was favored by the majority of the participants, and at the third general meeting, held in London on June 30, 1986, the members decided to set up the secretariat in Brussels.

The choice of the location of any European organization always carries political significance in European capitals, which shows what a long way we still have to go before achieving an integrated Europe. The French, who had initiated Eureka, would have liked to establish its seat in France, and they put Strasbourg forward. The West Germans agreed. Others suggested Geneva because of its obvious international character. Brussels did not meet with unanimous agreement. Some could see the advantages that might accrue from being located close to the seat of the EEC—advantages such as making use of the EEC's services and sharing its facilities. Others feared that the proximity to the EEC could cause confusion. Eventually Brussels was selected, but facilities separate from those of the EEC were established.

In the aftermath of the London meeting, a total of 72 projects were approved, calling for a total investment of about $2 billion. Thirty-seven additional projects were accepted at the Stockholm meeting on December 17, 1986. Among the first 72 projects, 17 were concerned with computer programming, 10 with new materials, 9 with biotechnology, 8 with robotics, 7 with microelectronics, 5 with town planning and transportation, 5 with environmental issues, 5 with production techniques, 5 with energy, 2 with communications, and 2 with oceanic studies. The projects that were approved in Stockholm involve various fields, such as the improvement of automatic oil platforms, the production of synthetic seeds, and the adjustment of powerful laser beams.

Some Specific Projects. It may be interesting to look at a few specific examples of Eureka-sponsored endeavors. The "Carmat 2,000" project teamed the French motorcar firm of Peugeot S.A. with the chemical firms of Imperial Chemical Industries (ICI) in the United Kingdom, Baden Aniline and Soda Factory (BASF) in West Germany, DSM in the Netherlands, Vitrotex in Italy, and Cristaleria Española in Spain. This project is estimated to cost $60 million and aims to develop a new conception for motorcar frames that would utilize new materials and reduce prices considerably. The "Paradi" project applies artificial intelligence to the development of a production and logistics management system that will meet schedules more efficiently, minimize inventory, optimize cash flows, and maximize human potential. Its budget is $30 million, and it links Aerospatiale in France, ABSY in Belgium, Aeritalia in Italy, Matrici in Spain, Brown Boveri in Switzerland and IKOSS in West Germany. The "ES II" (European Silicon Structure) project is oriented toward developing a European industrial infrastructure that will provide studies and realizations within the CMOS (complementary metal oxide semiconductor) integrated system by improving schedules and prices. This project, which should amount to $80 million, associates the Société Bull in France with Philips in the United States, Olivetti in Italy, and British Aerospace.

The financing of projects within the Eureka program is, of course, the crux of the matter. To give its partners an idea of the cost at the meeting where Eureka was created, the French government committed itself to investing an initial $115 million to get the program started. The West German government thought it might contribute up to $100 million. The British government was anxious to find means of financing outside the public sector. These issues were discussed anew in Stockholm, and it was decided that in addition to seeking financing for projects from industrialists, banks should assume a larger role. The Deutsche Bank has offered to summon together a "roundtable" to raise private capital to encourage European technological cooperation. The EEC could facilitate such operations by offering banks a guarantee for at least part of the money. Things now seem to be moving in the right direction, and European financial authorities are most likely to become increasingly interested in joint ventures such as those within the Eureka program.

Non-European Involvement. There is one other issue that cannot be avoided when discussing Eureka: Will American or Japanese firms be involved in the program, either directly or through their European subsidiaries? The same question has already been raised about the ESPRIT program. After a period of sharp opposition, opinions about non-European participation in ESPRIT have relaxed somewhat; thus, the International Business Machines Corporation (IBM) has become a partner in the program. I personally think that in this field it would be counterproductive to establish a very strict dogma. The Eureka program is intended to stimulate the revival of European industry, which is not a goal that American or Japanese firms have uppermost in their minds. However, this revival could in some cases result from agreements with non-European and multinational firms, especially when they have very important European subsidiaries. Such alliances will be interesting for Europe provided that the decision centers in these firms are neither essentially nor exclusively non-European.

An Ambitious Pragmatism: The Way Ahead

One could discuss many other topics directly related to Europe's technological revival. The subject of education and training is particularly important. Europe's languages and cultures are so numerous as to become impediments, but they could also be used to Europe's advantage. As education expands, it presents an opportunity for Europeans to start afresh with new methods that should be developed at a European-wide level. Some initiatives in education directly linked to the EEC are under way.

Europe must also commit itself to pragmatism as far as scientific research and technological development are concerned. The number of national governments results in a variety of elaborations as to how community cooperation should operate. Each European nation enjoys the idea that European-wide organizations benefit from what each one assumes to be the advantages of its own system. The result is the adding up of shortcomings. Nations must be willing to give up refined structures in order to stress the content of programs. It would be better to have programs without structures than to have structures without programs. However, Europeans should try not to exaggerate either tendency.

Another notion that is found in Europe as a result of its ancient cultural traditions is that science ought to rule over technology. Hence the natural tendency to believe that fundamental research will lead to success in applied research and development and that the purest science ought to come first in receiving the most careful attention. This tendency, however attractive, has become less and less realistic. There are very good reasons why basic research should develop in a technically superior country or region. But the reverse is much less obvious. Traditions or preconceptions explain why individuals in many European countries find it difficult to move from basic research to technology. Researchers have the same attitude as the nobility of the *Ancien Régime*: they must not step down. And money is not enough of an incentive to lure fundamental researchers into industrial research in sufficient numbers. The flow of knowledge, however, is linked with physical mobility. There is some hope that a lowering of geographical barriers will help to smooth out psychological barriers.

Europe's firm decision to develop highly advanced technological programs at home should have one final consequence—that of keeping the greatest number of Europe's best brains on the continent and of attracting many more from other countries. Should Europeans stop tackling the most difficult scientific and technical problems, the map of intelligence will quickly undergo important changes that will not particularly favor the continent. Fortunately Europe has not yet come to this point; it is even far from it. Still, to avoid scientific and technical impoverishment, Europe must respond to its problems with consistency, if not perfect order.

Whenever one of the ventures Europe has undertaken to encourage a technological revival meets with success, many people claim to be its father. Eureka, which now seems successful, enjoys a great number of supposed fathers. Yet there are dangers. Everybody is fonder of roses than of thorns, but thorns cannot always be meant for others. Each European country must come to terms with its own share of thorns. And we still have a great many that will probably prick us in the times to come.

Europe's scientific and technological ambition extends far beyond simple steps. It has already resulted in a true move forward for Europe. This advance will be useful, provided it is taken in a spirit of friendly agreement with the United States. Scientists and technicians are too well aware of fundamental values to be persuaded to pursue paths that would not provide them with the assurance of trans-Atlantic harmony.

Carlo De Benedetti

Europe's New Role in a Global Market

In the late 1960s, as the period of rapid economic growth that had characterized postwar European economic development was drawing to a close, it became apparent that a technological gap existed between Europe and the United States. Because their economies were lagging behind the United States, particularly in the high-tech realm, the major European governments initiated large-scale programs to promote their national high-tech industries, especially in energy, aerospace and computers.

Unfortunately, the various European policies did not succeed in closing the high-tech gap with the United States due to mistakes in identifying problems and inadequate action. Too much emphasis was placed on basic research, too little on the channels for bringing scientific research from the laboratories to the marketplace and to concrete industrial initiatives. Many of Europe's scientists and engineers were attracted to the United States; other resources were wasted by fragmentation and duplication at the national level. In the high-tech sectors the "national champions" (i.e., companies favored by their respective national governments) never reached a level of real competitiveness due to the overly protected environment in which they operated.

Now, after the crises of the 1970s, when reducing stagflation was the main concern of all the European countries, the problem of the trans-Atlantic technological gap has reemerged—but with Japan also playing an increasingly important role. Recent discussions on the U.S. Strategic Defense Initiative and Europe's "Eureka" program (promoting cooperation among European industries in the high-technology realm) have further highlighted the problem. But undoubtedly the main reason for Europe's renewed interest is the prospect of a new industrial cycle in which high technology will play the crucial role.

The long economic crises widened the gap between Europe and the two other major industrial regions, not only in terms of technology but even more so in overall industrial competitiveness. This gap—measured in terms of higher unemployment rates, higher costs, lower production capacity, and lower structural competitiveness in high-tech sectors—has created great concern about the future of Europe's industry and economy. At the same time, it has created a new consciousness which, through a new entrepreneurial wave, is pervading many European companies and countries.

It is more and more widely understood that Europe's weaknesses are a result of excessive market fragmentation, national protectionism, inefficient and costly public intervention, and a reduced capacity to compete in the new industrial cycle. New external conditions that favor a new entrepreneurial climate and a reduction in constraints are developing. The simultaneous fall in oil prices and the weakening of the U.S. dollar in 1985–1986 have had positive effects in terms of a reduction in inflationary pressures and in external trade deficits in most European countries, thus strengthening the new cycle of transformation that has been under way since the early 1980s. It seems possible today to stimulate a quicker response from European industry to the new challenges of a more rapid integration of the European internal market and of a growing entrepreneurial involvement in joint research efforts related to specific programs such as ESPRIT (the European Strategic Program for R&D in Information Technology), RACE (R&D in Advanced Communications-technology in Europe), and Eureka.

The New Industrial Cycle—Opportunities for Europe

The major factor promoting change is without doubt awareness that a new industrial cycle is starting based on new "soft" technologies—e.g., information technologies, knowledge technologies—that do not require heavy fixed investment but more human capital, education, research and development (R&D), adaptability, high value-added services, and networks.

Opportunities for Europe exist in this new cycle because Europe has a widely spread scientific and educational system; capabilities, derived from deep cultural roots, to generate new ideas; and a tendency toward a synthetic rather than analytical approach. (The ability to synthesize and generalize will be decisive in the new cycle.)

For these reasons, it is widely believed that Europe can play a new role in the coming industrial cycle, but also that investments are urgently needed in the intangible resources of science, education, information, communications, and entrepreneurship, and that these resources must be utilized in the best way. New synergistic cooperation is necessary between private entrepreneurs and government programs, the common aim of which should be growth and the reorientation of European industries to new industrial prospects. It is well understood that the new industrial cycle is not compatible with present national boundaries. What is required is not only a true European-wide approach but a more truly global approach. While the creation of a unified European market is therefore a must, it is not enough.

Stimulated by the expanding presence of new technologies and by the extraordinary growth of communications (both physical transportation and information transfer), the world is becoming a "global village." This process is being accelerated by the development of new kinds of interdependencies and alliances among industrial companies and by increasing exchanges of technology worldwide through joint ventures, cooperative research, and new deals that consider technology a commodity that can be easily acquired where it is available.

The current globalization is not limited to the industrial areas but tends also to modify relations with other, less industrialized areas. This can influence governments to move faster toward more rational international cooperation and coordination of economic policies on the basis of common objectives.

A return to growth that takes the best advantage of the opportunities offered by new technology is possible only if concrete international integration can be achieved. Because of the persistence of critical structural problems such as unemployment and unequal development within countries and between northern and southern regions, Europe can and must play an active role in the new cycle, performing as a catalyst in the process of international integration and globalization that will benefit all areas.

The opportunities created by technological change must not be lost. Entrepreneurs in Europe, as in the United States and Japan, understand that the new cycle requires full acceptance of the new challenges: this means abandoning pure conservatism, limiting protectionism and self-defense measures, opening markets, and achieving real integration on a worldwide level. Companies are now less and less limited by national boundaries; they can overcome technological gaps at the national level through the creation of cross-border technological networks.

European entrepreneurs know that the real technological gap today is the inability to participate on an equal basis in international bargains. The partnership created by Olivetti with American Telephone and Telegraph demonstrates better than all the books on the technological gap that when an agreement between two enterprises of totally different dimensions and cultures is possible and when Europeans are able to create the conditions for effective cooperation on an equal basis, no gap exists between European and American high technology.

Governments in Europe, and also in the other two areas, have a choice today. The positive decision would be to support and favor the natural path followed by industrial enterprises toward growing integration and coordination at European and international levels, thus accelerating a new phase of growth, creating new jobs and enterprises, and reducing the technological and competitive differences between companies. The alternative would be to add further constraints and obstacles to the natural tendencies in order to protect some limited national advantages, but at the cost of hindering the diffusion of new technologies, limiting new job creation, and, in the longer term, reducing their competitiveness.

The European Technology Gap

Comparisons between economic development in Europe and the United States tend to concentrate on the indicators revealing a technological gap. These indicators should be handled with caution, as should any conclusions drawn from an analysis of them.

Many indicators are available today to measure the different levels of scientific and technological development in the industrial countries: for example, per-capita R&D expenditure, number of researchers, number of patents, number of citations in scientific publications, etc. But they are generally over-aggregated and therefore lead to over-generalized or misleading conclusions. This makes it impossible to identify those technological areas in which Europe is ahead of or on a par with the other areas.

While emphasis is given to the overall gap as shown by the aggregate indicators, the areas in which Europe has an important and, in some cases leading, technological position are frequently neglected: for example, nuclear energy, aerospace development, transportation systems, biotechnology, flexible manufacturing, and some areas of information technology and electronics. Moreover, general comparisons between Europe on the one hand and the United States and Japan on the other are superficial. The analysis should evaluate the relative position of each area in specific technological sectors and trajectories.

New technologies are spreading more rapidly than previous technologies, and in a completely different manner. Static analyses therefore have less importance in an evaluation of technological gaps between countries. The development and diffusion of new technologies depend not just on the amount of R&D but, above all, on the circulation of innovative ideas and information (in the area of "soft" technologies, in particular) and on skilled labor mobility. Internationalization and standardization of products and markets are accelerating imitative innovation and the diffusion of new technologies. Therefore, no technological advantage or gap is permanent, and relative positions can change in a short space of time. For these reasons, I believe that the importance attached to technology in explaining Europe's competitive gap is exaggerated.

It is frequently stated that Europe's lag in advanced technology is due to insufficient R&D spending, but this is not true. European R&D spending is almost double that of Japan and about two-thirds that of the United States. The problem in Europe is that in many cases R&D initiatives are duplicated, so that resources are unproductively utilized, or research is mainly carried out by public laboratories and universities without any link-up with industry.

It is probably impossible to make an exact calculation of the sums spent by governments in France, West Germany and the United Kingdom to keep their national mainframe computer companies alive. But the figures certainly run into many billions of dollars, and the end result is well-known: today these 3 national champions have together less than 5 percent of the world mainframe market, even though they now sell mainly Japanese or American products, while IBM alone has over 70 percent.

National champion policies in high-tech sectors have failed to promote a competitive European industry, because their fundamental premise is incorrect. Government protection has very often slowed down the innovation process: the protected company feels less need to achieve greater competitiveness by investing in new technologies, developing new products, entering new markets, or reaching new international alliances. The unprotected company is heavily penalized in its home market due to a lack of free market conditions, and its growth is seriously handicapped. In telecommunications, for example, PTT (public telephone and telegraph) monopolies protect national industries almost everywhere in Europe, sheltering them from more innovative competitors.

Europeans are beginning to understand just how expensive protectionist, monopolistic policies are in terms of delayed development of new telecommunications services, high tariffs, inefficiency, and wasted resources. But almost no one, not even the "deregulated" British Telecom, has the courage to sail out into the open sea of the free market and true competition among European and non-European companies. Fragmented supply is also indicative of the divisions that exist among the European markets—divisions created by historical, cultural, ethnic, linguistic and political factors but accentuated by the economic and industrial policies of the individual countries.

In the last 20 years, the proportion of high-tech products in European Economic Community (EEC) exports of manufactured goods has remained unchanged at about 24 percent. During the same period, U.S. specialization in advanced technology rose from 29 percent to 33 percent, and Japanese specialization from 16 percent to 38 percent. A gap exists because Europe has not capitalized on the resources it has invested in R&D and has not generated a proportionate level of new products. It exists because Europe was, and still is, lagging behind in the development of a larger unified market and an innovative environment. Inadequate growth, high unemployment, higher costs, lower industrial capacity, and falling competitiveness in high technology are the most evident symptoms of Europe's malaise. They worsened during the crises of the 1970s, because Europe, unlike the United States and Japan, failed to make an integrated response.

The problem lies not, as it appeared during the debate on technology in the late 1960s, in insufficient research investment but in an inadequate ability to translate the results of research into new ventures and new products; in the duplication of efforts because of market fragmentation; and in an inability to achieve sufficient economies of scale. I shall mention just a few significant indicators of Europe's competitive delays. My intention is not to fuel a sterile mood of "Euro-pessimism" but, on the contrary, to show that it is possible for Europe to reverse its present trend and to participate actively in the new cycle of growth.

In the three-year period from 1983 through 1985, Europe recorded overall growth of 6 percent, compared with approximately 14 percent in the United States and Japan. Production investments rose in real terms by less that 10 percent, very far from the 37 percent achieved in the United States during the same period. Slower development and fewer investments have resulted in less innovation and a lower capacity to adapt to new markets.

Europe's growth was no better in qualitative terms. Average annual inflation in the EEC from 1983 through 1985 was 6.5 percent, significantly higher than the 3.8 percent in the United States and 2.2 percent in Japan. Furthermore, European unemployment rose to over 11 percent of the labor force, while it was less than 7.5 percent in the United States and 2.5 percent in Japan. In 1975, Europe had 8 million registered unemployed workers. In 1986, the figure exceeds 19 million and continues to rise, albeit slowly.

Over 35 percent of the unemployed have been out of work for more than a year. Many have been without jobs for over two years—22 percent of the total unemployed in each of the United

Kingdom and France, 32 percent in Spain, and 49 percent in Belgium. People who have been excluded from the labor market for such long periods are not only highly demoralized but economically hard-up. Their work skills deteriorate, and a vicious circle is created as their chances of employment decline.

A more detailed analysis would also show that the problem is aggravated by the fact that unemployment is heaviest among young people looking for their first job. Almost 44 percent of Europe's 19 million unemployed are between the ages of 15 and 24. In France, the youth unemployment rate has doubled from 15 percent to 30 percent since 1980. This means that the average age of employed workers tends to rise, so that the stimulus for renewal provided by new generations within companies and institutional structures is disappearing.

This situation, caused by the extreme rigidity of the labor market, is particularly serious because it comes at a time of important changes in technologies and in work skills. In Europe, only 10 percent of the work force changes jobs during the year, compared with 28 percent in the United States.*

This rigidity extends beyond the labor market, however. The industrial renewal rate, measured as the percentage of new manufacturing companies to existing companies, is less than 2.5 percent in the major European countries, in comparison to 4 percent in the United States and 7.5 percent in Japan. In other words, industrial renewal in Europe is only half that of the United States and one-third that of Japan. But the problem in Europe is not an inability to create new companies, at least if the entire economic system is considered and not just the manufacturing sector. In fact, the 600,000 new companies formed in the United States in 1985 are comparable, in proportional terms, to the 170,000 created in the United Kingdom and the 150,000 in West Germany. The difficulty lies in enabling the new companies to grow.

**Editor's note.* For a further discussion of unemployment and labor rigidity in Western Europe, see *Unemployment and Growth in the Western Economies,* Andrew J. Pierre, editor, New York: Council on Foreign Relations, 1984.

American experience suggests that economic growth based on entrepreneurial dynamism is accompanied by a high mortality rate among new ventures. U.S. companies with less than 20 employees have less than a 40 percent probability of surviving for more than four years; but those that do survive often become medium or large companies. In other words, they are bound to succeed.

In Europe, on the other hand, the failure of a new entrepreneurial venture is not by any means considered a normal occurrence. It is a social stigma: the entrepreneur is denied credit, his credibility evaporates as do his chances of forming a new venture. It is therefore not surprising that new (and old) companies are not greatly inclined to take risks and that small companies remain so, not growing beyond certain levels. This is true despite the fact that one of the ingredients for the development of innovation is the willingness to take risks.

Of the many other factors cited as obstacles to the development of Europe's entrepreneurial capabilities, one of the most prominent is the absence of a truly unified market, similar to that of the United States, which can provide a sound base for expansion worldwide. Another obstacle is Europe's lack of financial institutions and new financial tools at a European level. The tendency to safeguard "lame ducks" has been detrimental as well. Companies in declining sectors have not been shut down in Europe because no one has had the confidence to create new enterprises in the emerging sectors.

In the 1970s, strong union power and the abnormal growth of the welfare state created a no-growth culture in which entrepreneurial initiatives were not encouraged. The "development and risk culture" was replaced by the "public assistance culture." The resulting weakness in European competitiveness slowed down industrial restructuring and has made it difficult to cut down high unemployment in certain European countries. It fostered protectionism, which further reduced competitive capacity in the new technologies.

For this reason, I believe that Europe's weakness in the high-tech realm and in industrial competitiveness is not a matter that concerns Europe alone. Europe's weakness, like any U.S. or Japanese competitive weakness, has a negative impact on all industrial countries at the world level. Imbalances created in one market have a detrimental effect on other markets at the international level. In high technology areas, permanent trade surpluses or deficits set off chain reactions that provoke varying degrees of protectionism. Protectionism at the national level is an obstacle to integration at the company level.

A recent survey of more than 100 top European corporate executives conducted by *The Wall Street Journal*/Booz-Allen & Hamilton Inc.[1] shows that the majority of respondents think the lack of market integration is the major factor in Europe's inability to maintain a competitive edge against the United States and Japan. This means higher costs, duplication of investments, lack of competition due to de facto industry cartels and monopolies, different health and safety specifications, proliferation of standards, and costly national industry purchasing practices.

The survey provides a ranking of the sources of extra costs in doing business across European borders:

- different specifications set by national laws—42 percent
- sheer cost of paperwork—25 percent
- extra warehousing and inventory costs due to different product standards—20 percent
- different specifications set by national industry groups—19 percent
- cost of complying with "country of origin" rules—18 percent
- extra warehousing and inventory costs due to distance and logistical factors—18 percent
- distance and lack of adequate international transport systems—14 percent
- different consumer preferences—13 percent

Seventy-five percent of the panel believe the lack of a unified European market leads to higher R&D costs due to the unnecessary duplication of R&D laboratories, extra costs to meet differ-

---

[1]See *The Wall Street Journal*, March 6, 1986.

ent national technical specifications, and costs incurred because of delays in getting products to the market. But the same survey indicates that the majority of respondents are optimistic that Europe can modify this situation: industry is becoming more aware of the fact that if advantage is not taken of the opportunities presented by technology, European enterprises will not survive the coming industrial restructuring at the international level.

A "change or die" approach is spreading among European companies with the awareness that the leading role in the phase of change has to be played by industry itself. Some examination of the changing economic scenario is useful for an understanding of the new European entrepreneurial spirit.

A New Scenario for Europe and the Industrial World

A study of long-term modern economic history reveals the presence (or absence) of technological progress behind many cyclical fluctuations, behind the success or economic decline of a nation or company. In the past, technological innovation progressed as a series of fairly well-defined waves. Each wave was distinguished by the introduction of several new basic technologies which, though not necessarily linked with one another, provided the raw material for a cascade of innovations that radically transformed the technological and economic scenario. This was the case with the coal-steel-railways wave, and the oil-automobile-aeronautics-organic chemistry wave, and so on.

A new wave is now building up based on information technologies-microelectronics-biotechnology-new materials. Many observers believe that this wave will have a greater impact than its predecessors for at least two reasons. First, never before has there been a convergence of new technological developments toward the conquest of new frontiers, not just in specific industrial sectors, but across the entire production system. (A term often used for the new technologies that have this extraordinary capacity for diffusion is "infratechnologies.") This is leading to radical changes in products and in production processes, while the logic of innovation and of renewal common to all economic activities is drawing the different sectors closer together. New opportunities are thus being created for exchanges among the different technological areas and, although it is impossible today to imagine the level of innovative synergy that will be achieved, this will give still greater vigor to the new technological wave. Second, the new technologies will not only expand physical resources, as in previous cycles, but they will above all expand man's immaterial resources—i.e., intelligence and knowledge.

Technological progress might be described as a series of parabolas, each one higher than its predecessor. Those who remain on the old parabola can still progress but only within certain limits and for a limited period of time. When they reach the peak of the curve, the only way they can go is down. The new technological curve—or wave—is now building up. Only those companies and countries able to mount and master it can hope to be active participants in a new phase of economic growth, developing all those derived innovations whose technological value is perhaps limited in relation to the basic innovation, but whose commercial value is high.

On the other hand, strategies based on the simple renewal of traditional activities, with marginal use of new technologies, are bound to fail over the long term, even though they may be profitable in the short run. New companies are formed to take prompt advantage of the opportunities created by a new product or by a new idea, and at the same time old companies that are unable to renew themselves are pushed out of the market: this radical restructuring is occurring in all sectors. This turbulent process obliges companies to constantly renew their technologies, products, production processes and also their market operations. In the circumstances, no position is secure and nothing can be taken for granted. This applies to enterprises, but it could also be applied to countries.

These considerations on technology lead us to another important factor of the new cycle—the internationalization of economic systems. In the two-year period from 1984 through 1985, world growth was 8 percent while international trading volumes increased by 15–16 percent, an elasticity factor of nearly 2. This is important because it indicates that the protectionist phase that followed the oil shocks of the 1970s has been bypassed. From 1973 to 1983, world trade grew by 32 percent and world gross domestic product by 28 percent, with a very low elasticity factor of only 1.15. World trade has thus recovered and is once again—as in the 1960s—the locomotive of growth.

A major cause of the current internationalization is the exceptional rise in communication processes based on the development of rapid low-cost means of physical transport and advances in computer and communication technologies that enable information and knowledge to be transmitted at unprecedented speeds. The new telecommunications, in particular, play an important role. Satellites and new telematic services infinitely multiply the possibilities of transferring and acquiring information, ideas and knowledge. This has stimulated an extraordinary circulation of people, products and processes, and therefore of culture, tastes and attitudes.

The emergent cultural and economic process reduces distances in time and space, and transforms the world into a great village, where each advance can be immediately available to everyone. The new situation is clearly evident to those who work in high technology, but it will increasingly extend to all sectors and activities, from industry to banks and services in general. Only those who become "global operators" will be able to compete in this scenario.

The approach of individual companies to markets is also changing to reflect the new situation. In the past, companies operating at the international level took the classic multinational corporation as their reference model. Today, an international presence can be obtained not through direct foreign investment as in the past, but through the development of a complex network of alliances, joint ventures, or partnerships between large and small companies worldwide.

Access to technology can only be achieved in an international perspective today, for at least two reasons. First, the rate of technological change is so high that product life cycles have been considerably shortened, and research, design and engineering costs have increased. Second, radical changes in organization stimulate internationalization. For example, in manufacturing, the trend is toward a reduction of traditional vertical integration and a growing segmentation of production into phases. The separation and coordination of the manufacturing phases are made possible by information technology in assigning the various phases to different companies. The end product is very often a modular recomposition of various parts and components produced in different companies and different countries. In this way, the traditional relationships between small and large companies are changing: synergistic integration is possible among companies of different dimensions without leading to the technological or managerial subordination of the small firm to the large. Cooperation can be undertaken where the joint effort brings benefits to both partners.

Within these "constellations," where each company retains its managerial and decision-making autonomy, synergistic integration would be directly proportional to the degree to which the barrier of nationality can be broken down. As a result, although the end product would still have a national brand name, its components and added value would be increasingly less identifiable with a single nation.

As it becomes more difficult to analyze trade flows, the criteria that measure competitiveness at country level are changing. Export competitiveness cannot in fact be achieved by an industrial system simply through low production costs, an undervalued exchange rate or other forms of export subsidy; it depends above all on the ability of the industry in that system to participate actively in the international network of new industrial alliances and partnerships. In this situation, governments may appear to have fewer opportunities to pursue an active export policy, since they are to a certain extent unable to control the strategic decisions of supranational networks. In reality, however, the sovereignty of governments remains in creating the economic, technological and financial environment to enable national industry to become an active member of these networks.

Finally, another cause of change in the next few years will be the growing role of the services sector. This is not a new trend; what is new, perhaps, is the increasing importance of services in international trade. From 1969 to 1982, world exports of goods rose by 632 percent, while service exports (i.e., revenues from transportation, tourism, financial and professional services, etc.) rose by 685 percent. Services thus expanded to account for over 21 percent of world exports, and will tend to increase further.[2] Until recently, international competitiveness was a requirement for industry alone. But today no country can underestimate the importance of an efficient and competitive system of services. It would therefore be a mistake to analyze a production system's competitiveness solely on the basis of its industrial trade balance. Services, whose circulation will also be favored by the new telecommunications networks, will become an increasingly autonomous factor in competitiveness.

Europe's New Entrepreneurial Spirit

The opportunities offered by the new industrial cycle are now clearly understood by European entrepreneurs. Since the early 1980s Europe, at first slowly and then more rapidly, has been shaking itself free of the uncertainty and pessimism that some have defined as Euro-pessimism or "Eurosclerosis." This change of attitude originated essentially in the entrepreneurial world, but a new approach is also being adopted in national and EEC policies.

European enterprises are looking beyond national horizons to develop alliances and partnerships at European and international levels. Enterprises are placing greater confidence in their own capabilities and resources instead of depending on public subsidies. In many cases, the change has come from new generations of entrepreneurs, from companies that operate outside the more traditional industrialized areas.

An entrepreneurial breeze is blowing through Europe emanating from southern countries, like Italy, but pervading the whole continent. Investments in manufacturing automation, R&D, and technological innovation have made a strong recovery. European cooperation in R&D and production has been boosted by the introduction of the ESPRIT and RACE programs. These programs enable competitors to sit around the same table for the first time and to recognize that they have the same prob-

---

[2] *World Invisible Trade,* British Invisible Export Council, London, July 1984.

lems to solve. Although cooperation is developing in the traditional as well as the high-tech sectors, the major stimulus comes without doubt from the prospects opened up by the new technologies.

The development of cooperation and partnership among European industry depends upon avoiding the reproduction at the European level of the defects and limits of past autarchical national experiences. European entrepreneurs understand that the new industrial cycle does not allow for limited horizons but requires a global approach. European market unification is necessary, but it is not enough in the industrial cycle. For this reason, partnerships are multiplying on a real international basis.

European companies are no longer targets for takeovers by companies from other areas; on the contrary, they are operating with greater vitality in international markets and are developing innovative acquisitions and alliances. Unification is taking place at business and market levels from an increasingly common linguistic (English) and cultural base, leading toward a pragmatically unified Europe without raising the thorny question of political unification.

These new trends do not mean that the many large obstacles and difficulties, described earlier, have suddenly disappeared. But overcoming these obstacles now appears less difficult than in the 1970s, when Europe seemed to be heading down an endless dark tunnel. These changes and the new European entrepreneurial spirit must be taken into account by the entrepreneurs and politicians of the other major industrial areas, because effective progress at the world level today will come from a clear understanding of the different situations and from agreement on the common objectives to be achieved.

On the basis of my experience as an entrepreneur, I believe that Europe must pursue four fundamental objectives in order to reduce the entrepreneurial gap and to promote change:

1) the integration of European systems at all levels (including monetary and economic policies);
2) the development of a unified European capital market, with a common European currency, Eurostocks, and com-

mon tools to channel savings toward innovative companies and to give more transparency to financial investments;

3) the promotion of a decisive entrepreneurial approach, not just in companies, but in government and public services as well;
4) greater scientific development and utilization of technology based on closer cooperation among companies, universities, and public laboratories and on technological partnerships with U.S. and Japanese companies.

The *first* goal, integration of European systems, is a demanding one. Nonetheless the European nations must transform themselves into a united global competitor over the coming years if they want to be winners in international markets. Without a truly common market open to competition, and without a reasonable level of integration among financial, fiscal, judicial and social structures, the European nations are bound to lose ground technologically and strategically to the United States and Japan.

The community created by the Treaty of Rome is nearly 30 years old; it has made much progress and grown in size, but now it needs to make a rapid qualitative leap forward, because there is little time left to become part of the new cycle of development and technological renewal. This means that the European countries with the greater industrial resources must take the lead in refounding and integrating Europe.

The *second* goal that must be achieved is the development of a unified capital market, the wide use of the European currency unit (ECU), the widespread utilization of Eurostocks, the creation of European venture capital funds (in the fashion of Euroventures—a European network of Venture Capital Funds created by a group of large European companies). The renewal and restructuring of the European industrial system requires a major accumulation of new capital. Industry must have access to a high level of resources to launch a vigorous long-term investment cycle. Company profits and stock values improved almost everywhere in Europe from 1984 to 1986. But a cyclical rise in profitability is not enough. Companies must be more substantially recapitalized through greater inflows of new capital.

European private savings must be directed toward productive investments through closer links between savers and industrial investments, instead of being used to finance public debt and to shore up the operating losses of debt-ridden companies and public institutions. Savings must have a real dynamic counterpart capable of generating new wealth. This is only possible if they are invested in truly productive activities that create profit and employment, instead of being wasted by putting money into sectors and industries that are unlikely to have any growth. The right conditions and tools must be created to channel savings into industrial investments.

One condition is adequate remuneration for invested capital. In other words, companies must be in a position to be profitable. Next, a wide variety of possible investments must be available to the saver, who will base his choice on an objective evaluation of company profitability. This can only happen where there is complete market transparency, where the saver invests his money in entrepreneurially run companies and not in public institutions. Europe also needs new tools to channel savings toward companies, in particular toward innovative companies. The stock market is certainly a key tool. But although almost all the European stock markets have made great progress in the last two years, there is still a clear need for reform and modernization everywhere.

The many tools available on the sophisticated American financial market are certainly an important factor in the financing and successful growth of innovative and profitable companies. Recent experience in Italy also demonstrates the importance of financial innovation. In 1983–1984, new rulings were introduced regarding mutual investment funds. From the end of 1984 to the middle of 1986 more than 50 funds were created and have collected the equivalent of over $35 billion. An important portion of these funds has been invested in the stock market, which has made enormous progress. For their part, companies are making greater use of capital increases, thus unblocking a stunted market.

The *third* area on which Europe's recovery depends is entrepreneurialism, which must be promoted throughout the European system in the management, not only of companies, but also of government and public services.

For many years the culture prevailing in Europe was based on guaranteeism, conservatism, protectionism—the welfare state at any cost. As a result, many companies, which would otherwise have disappeared from the market, were able to survive for years through public subsidies and contributions, preventing other healthier concerns from establishing themselves. Under state ownership and protection, many enterprises were managed by bureaucrats, called upon to run in a conservative manner companies whose international competitiveness constantly declined.

A new wave of privatization and deregulation is spreading throughout Europe, but much remains to be done. The entrepreneurial approach needs to be promoted and motivated through adequate legislation. There are numerous examples of fiscal, financial or trade union laws in Europe that discourage both the formation of new entrepreneurial ventures and the development of small companies beyond certain thresholds.

The same entrepreneurial approach should also be adopted in the public sector, which needs to be renewed no less than industry. Models inspired by the "assistance economy" have absorbed enormous resources and created public deficits, which greatly limit economic policy decisions. More entrepreneurialism in the public sector does not, however, mean less government. If anything, it means less bureaucracy, less waste, less conservationism. It means an opportunity to reduce costs, to improve public services and infrastructural investments, and to amortize the social cost of painful industrial restructurings.

The *fourth* direction in which Europe must move is the concentration of R&D resources and cooperation in high-technology areas (mainly in information technology, which today represents the vital element for growth in all sectors). As mentioned above, Europe's slow progress in this realm is caused only to a small extent by insufficient investment in R&D. Europeans must understand that innovation is less and less the product of individual genius and increasingly the outcome of a system approach, a system that includes large and small companies, universities, and private and public research laboratories. The integration of these various parts should be ensured by adequate legislation and by public procurement as a stimulus for technological innovation.

Europe must learn from changes in antitrust legislation in the United States. These moves have promoted an increase in alliances among advanced companies, particularly in R&D cooperation. European governments should be inspired by these examples to launch a new common R&D strategy promoting cooperation among companies of different countries. The ESPRIT projects and the more recent Eureka program are going in the right direction, but they are still too limited in terms of financial resources and real willingness to coordinate R&D efforts.

Conclusions

If the debate on the technological gap existing between Europe and the United States and Japan is to be constructive, it must analyze the factors which, in each of the three regions, block or limit synergistic participation in the new cycle based on the intensive application of new technologies.

To exploit fully the opportunities offered by the new technological "waves," the three major industrial areas must establish common objectives for global action:

1) the free international exchange of skilled labor, technologies, information, products, and entrepreneurial initiatives;
2) real reduction of protectionism to open up world markets;
3) development of entrepreneurialism and alliances among companies; only supranational enterprise networks can reduce technological and entrepreneurial differences between countries, and create greater worldwide opportunities for a new phase of balanced economic growth.

A "new purpose" is spreading among European entrepreneurs to achieve unrestricted development open to growing internationalization. The revitalized entrepreneurial spirit now emerging in Europe must be strengthened. There is greater impetus in Europe for true technological and entrepreneurial cooperation, both within Europe itself, as well as with the United

States and Japan. The partnerships, alliances and new forms of interdependence being developed among companies enable frontiers to be overcome, the stimulus of new technology to be transferred to the entire world industrial system, and the technological and industrial gaps between areas to be reduced. The cultural delays that still persist among those institutions and social groups that look to the past rather than to the future must be prevented from slowing down this process, in Europe as well as in the United States and Japan.

Disalignment is detrimental to all areas, whatever their individual rates of growth. The natural trends emerging today must be encouraged through appropriate economic policies that are designed to integrate rather than differentiate the various systems and to foster real technological and organizational innovation as a means of stimulating the worldwide diffusion of true free-market capitalism

Protectionism and all the barriers to greater competitiveness must be eliminated. The new technologies provide an extraordinary opportunity to reduce disaligned growth, not just in the industrialized areas but also between these areas and the less developed countries. It should not be forgotten that the solution to international security problems is strictly related to the capacity to create and spread growth and wealth worldwide.

The opportunity must not be wasted. Everyone must closely cooperate to achieve a cultural "leap" forward, so that the old models inherited from the crises of the 1970s can be permanently abandoned, and new models open to both quantitative and qualitative growth and true internationalization can be constructed. The new cycle is centered to a far greater extent than any previous phase on man and on the growth of his intelligence, his knowledge, and his culture. The technological challenge is not just a matter of economic or technical resources; it is above all a matter of culture, in which the true values of a civilization come into play and where Europe's great historical and cultural heritage can play a decisive role.

Keichi Oshima

The High Technology Gap: A View from Japan

The technology gap between Western Europe on the one hand and the United States and Japan on the other increasingly has become a major issue in the international debate. Yet, the argument about the existence of a technology gap is not new. In fact, in the 1960s the technology gap between the United States and other countries in the Organization for Economic Cooperation and Development became a dominant topic in the OECD's Committee on Science and Technology Policy when it examined science and technology as a means of achieving economic growth.

Unfortunately, the term "technology gap" is often used ambiguously—without being placed in a clear context—despite the fact that the nature, composition and policy implications of the gap differ substantially from context to context. For example, the implications of a technology gap in the area of trade or, more generally, in the economic realm are not necessarily the same as the implications of a gap in the military realm. Nonetheless, the basic elements comprising the technology gap in both these areas do not differ to a great degree.

Any examination of the technological competition between Japan and Western Europe must take into consideration three premises that serve to underline Japan's unique comparative position. First, Japan's widening of the technology gap vis-à-vis Europe during the post-World War II era was an unexpected by-product of Japan's more explicit effort to narrow its gap with the United States. Second, the process of Japan's drive to catch up with the West is important, because it was due to the perception that Japan was lagging that the term technology gap came to be used as a tool to encourage the development of Japanese industry. Third, as mentioned above, discussion of the technology gap should be specific in its context, since it has many facets and different implications—e.g., political, economic, etc. Therefore, no examination of the subject should fail to deal with current technopolitical problems, such as technology transfer, high-technology trade, and the role of international technological cooperation in the Western Alliance.

Given these premises, I will attempt to draw a basic picture of the technology gap between Japan and Europe in a broad context that includes the United States. I will present some basic facts about the technology involved and develop the three premises mentioned above. I will also discuss the process by which Japan used technology as a major thrust to gain a competitive edge in its modernization process. Finally, I will engage in a full discussion of how Japan presently views the technology gap vis-à-vis Europe.

Different Components of the Technology Gap

In order to clarify the issues, it is necessary to address the ambiguity of the term technology gap by defining science and technology in terms of objectives and impacts. There are four key dimensions—the intellectual, the economic, the political and the social—in the relationship between national policy and scientific and technological activities.

The intellectual dimension of science and technology deals with the creation of a body of knowledge that is man's common asset, because it is derived from man's intellectual activities. On the other hand, the economic dimension treats science and technology as a means of attaining economic objectives. In the same manner, the political dimension sees science and technology as representative of a nation's power in the international political arena. Viewed from these latter two perspectives, science and technology are a means to an end. Therefore, they encompass other elements such as the industrial power, research and development capability, military might, etc., of a particular nation. The social dimension of science and technology focuses on the impact made on society by science and technology. It thus denotes public acceptance of technology and its integration into art and culture. The "Luddites," for example, were a social group that rejected scientific and technological advances.

The basic reason that the technology issue has gained a central position in the recent debate among industrialized nations seems to stem from the fact that there has been a gradual shift in the international importance of science and technology from the second (i.e., the economic) to the third (i.e., the political) dimension. The technology gap has become a particularly heated issue among policymakers in Western Europe, the United States, and Japan, because the so-called technology intensive or high-technology industries have become a crucial ingredient in a nation's performance in international trade and economic activities. This is evidenced by the fact that the technology gap between Japan and Western Europe is perceived as a decisive factor in inducing unemployment and trade deficits in the latter. Recently, this impact has gone beyond the economic dimension and is recognized as having a direct impact on social stability and military affairs, thus threatening West European security vis-à-vis the Eastern bloc.

It is important to note, however, that technology itself is transferable in various forms. In fact, technological cooperation and transfers have occurred naturally and extensively at different levels, such as among firms, groups of industries, national agencies, and governments. Therefore, the lack of a fundamental technology or certain components of technology does not automatically imply a technology gap between two nations. If a country lacks in a particular technology, it can buy or obtain it through a scheme of bilateral technological cooperation. However, if technology is seen as a total system or process involving the whole innovation cycle (from research and development to production to marketing to practical use), what is necessary to realize a nation's objectives, whether they be civilian or military or both, is more than a simple transfer of specific component technologies.

Discussion about the technology gap between industrially advanced countries is now centered on the gap between Western Europe on the one hand and the United States and Japan on the other. Such a "gap" is often measured in terms of macro output variables such as the trade balance in technology-intensive products, numbers of patents, technology trades, and the like. However, if we seek to assess the economic and political implications of the gap, including the policy measures needed to ameliorate the situation, technology has to be treated as a total system. Then one must take into account broader input variables such as capital flow, management skills, quality control, labor/management response to productivity change, and even public acceptance.

In other words, the present issues cannot be argued by merely comparing the levels of various technologies, but must be looked at from the more fundamental standpoint of a nation's capability to manage the outcome of recent, dramatic technological innovations as a means of achieving socio-economic goals and security. The comparative ability of a national socio-economic system to adapt to structural changes induced by technological innovation will be an especially dominant factor in the discussion of international political issues.

Japan's Technology Gap: History and Perceptions

The Meiji Restoration. It seems imperative to start the discussion of the technology gap between Japan and Europe with a history of Japan's development beginning with the Meiji Restoration in 1868. Equally important is a review of how Japan endeavored to narrow the technology gap, first with Western Europe and then, after World War II, with the United States. Such an examination inevitably must touch upon Japan's perception of the technology gap and of industrial technology itself.

The unique pattern of Japan's development can be found in how the country viewed technology during the modernization process. After 300 years of closure of the country, the Meiji started to reform the whole political and social system, patterning it on the Western countries. Such Westernization was regarded as the most important means of establishing national power so that Japan could catch up with countries abroad industrially as well as militarily. Two slogans at that time, "Western skill with Japanese spirit" and "wealth and military power," symbolized this.

The build-up of national technological capacity by the introduction of Western technology was regarded as an integral part of the Westernization process from the outset. Reforms were taken by the government to upgrade education; transfer technology from abroad; and encourage indigenous research and development, commercialization, diffusion of technology, and product innovation. This integrated or deductive approach of Japanese science and technology policy was somehow different from the inductive approach of the Western countries. The focus in the West was on new inventions that were the result of research activities that naturally led to application and then to commercialization and diffusion. In Japan, the whole process came at once. This was due to the fact that Japan was a latecomer so that government and industry alike strove to catch up with the West. It is important to note that Japan's technology policy was incorporated in the nation's industrial development and military enhancement policies.

Clearly, the technology gap between Japan and the Western countries in both industrial and military technology was given special attention at the outset of the Meiji era. In order to fill this gap and move closer to the major world powers, the government implemented several key policies. It established engineering schools at university level, first under the Ministry of Industry and later under the Ministry of Education, which served as major sources of technological manpower for industry and government. Selected young students were sent abroad for study and training in science and technology. It is interesting to note that higher education in science and technology was strongly oriented toward practical objectives and that there was virtually no brain-drain since almost all students returned to Japan. These students have become not only professors or government officials, but many of them have played roles as leading engineers and innovators in industry. It is fair to say that from the start the practical application of science and technology was the predominant objective in Japan.

Major heavy industries like iron and steel, machinery, shipbuilding, chemicals, and coal mining were developed at government initiative to establish a nation of wealth and military power. These were national enterprises that allowed the participation of private groups. Later they were transferred entirely to the private sector, mainly to the major financial groups (or "Zaibatsu"), and became their industrial core. Naturally, the development of technology in heavy industries was aimed at strengthening national power, both militarily and economically.

Throughout Japan's history, technological innovation in industry has been achieved thanks to close ties between the government and pioneering Japanese enterprises. On the other hand, even in the early days of industrialization, many small, private, traditional industries adopted new technologies from abroad on their own initiative. Such entrepreneurship has been a continuous major driving force behind Japanese technological development. Sakishi Toyota, who introduced the electric motor to traditional weaving machines, is an example of one of these early entrepreneurs. His company grew into what is known today as the Toyota Motor Group. Many light industries such as those manufacturing matches, toothbrushes, shoes, glassware, soap and buttons adopted foreign technology as a means of modernizing their production and products. This private initiative was backed up with government assistance (e.g., technological help from local public laboratories) that encouraged the modernization.

The slogans "catch-up with the West" and "narrow the technology gap" were the dominant themes in the Japanese industrial world during the Meiji Restoration. The technology gap was used as a prime mover to modernize and industrialize Japan so it could become a world power. This national preoccupation was widespread not only among government and industrial elites but also among those at the lower end of the social structure.

Post World War II Era. This "narrowing-the-gap consciousness" persisted until the end of World War II. However, the early postwar era marked a conspicuous difference in many ways. First, Japan lost its independence in military power entirely and the development of military technology was banned, which led Japan to concentrate on civilian technology. Second, in the civilian realm, Japan became heavily dependent on the importation of technology mainly from the United States. Thus, Japan's prime concern during the postwar period was its gap with the United States in civilian technology.

After World War II, Japanese industry was completely destroyed; there was a tremendous technological gap between

Japan and the West, particularly vis-à-vis the United States. The productivity of Japanese industry was said to be five to ten percent that of the United States at the start of reconstruction in the late 1940s. Therefore the major postwar effort was directed at the recovery of Japanese industry under the various constraints imposed by the U.S. occupation forces.

However, in 1950, at the end of the occupation, the Japanese government started the so-called heavy (e.g., iron and steel) machinery and chemical industrialization. This narrowed the policy focus, transforming Japanese industry into a technologically advanced structure with enhanced competitiveness in the international trade market and enabling Japanese industry to catch up with industries in the West. Firms were encouraged to close the technology gap by importing foreign technology. From 1950 to 1959, technology imports amounted to more than 1,000 cases, and as Table I indicates, 64 percent of these were from the United States. In addition, in 1949 the Japanese government enacted the Industrial Standard Law for Japanese Industrial Standards, and in 1950 it introduced quality control modeled on the U.S. example in order to make Japanese products more competitive.

In the pre-war era, Japan had sought foreign advanced technology from not only the United States but European countries as well. However, its defeat in the "Pacific War" made Japan totally dependent on the United States and hence widened its distance from European countries. In fact, the majority of students who wanted to study abroad chose to go to the United States (in contrast with the pre-war period when Europe was the predominant choice); and Japanese industries cooperated more closely with U.S. universities than they did with European ones. Literally, Japan was incorporated in a "trans-Pacific technological alliance," in which the United States exerted hegemonic influence and thus set the common standards. At the same time, Japan's development of military technology ceased or became "passive," and its defense was incorporated into the U.S. security system under the 1952 U.S.-Japan Security Pact. Today, Japan does manufacture weapons for its Self-Defense Forces, in part through co-production arrangements with the United States, but it is not an exporter of arms.

TABLE I
Introduction of Foreign Technology to Japan (1950–1959)

Country	Percentage of Imports
U.S.A.	64.6
Switzerland	7.9
West Germany	6.9
U.K.	3.3
France	3.3
Netherlands	3.0
Italy	2.4
Canada	2.3
Sweden	2.0
Others	4.2

Source: White Paper on Science and Technology, Science and Technology Agency, Tokyo, 1979, p. 22.

As the result of such structural changes, postwar Japan concentrated its financial resources and best technical manpower in the civilian sector. Manpower and the best pre-war military technology was transferred to civilian use and contributed to enhancing the competitiveness of Japanese industry. This became apparent when Japan developed its camera and optical instruments technology (both by-products of naval research) and introduced the "Shinkan-sen" bullet train (which was developed by military aeronautic engineers). In other words, the best technicians made TV sets or washing machines in Japan, whereas that same class of people in Western nations were engaged in more sophisticated areas of research and development (R&D) such as aerospace or the military.

During this period broad government measures to implement industrial policy, such as tax benefits, financial assistance for investors, and the supplying of qualified manpower played an important role in establishing the economic and social environment needed to promote industrial innovation in Japan. However, it must be noted that the main actors and driving force came from the private sector. Severe competition in the domestic consumer market promoted rapid innovation and improvement in the quality of Japanese products. New ventures, such as Sony and Honda emerged from such competition with their own innovative technological capacity and established themselves as leading companies in the field without government assistance.

Japan re-emerged from the devastation of the war at a miraculous speed to become a powerful contender, but exclusively in the civilian sector. By the end of the 1960s Japan achieved significant success in fulfilling its goal of transforming its industrial base into one comprised of mostly heavy industries such as machinery and chemicals during a period of high economic growth not only in Japan but throughout the world. Innovation in industrial technology, especially production technology, was promoted, and the technology gap with the United States in the civilian sector was narrowed. Needless to say, Japan had fallen behind in military technology and its related fields. While the issue of military technology has been gradually introduced in discussions on overall U.S.-Japan relations, this has been primarily in the context of the application of Japan's civilian technology to the military realm.

The ratio between public and private initiative in Japan has stayed constant since the postwar era and is reflected in R&D statistics, which show that only 25 percent of Japan's R&D expenditure is undertaken by the government and 75 percent by the private sector. This is compared to government expenditure of 58 percent in France, 46 percent in the United States and 42 percent in the Federal Republic of Germany (see Table II).

The Oil Crises. The oil crises brought on a completely new situation. First, the 1973 oil crisis was a serious blow to Japan because of its heavy dependence on imported oil. Oil comprised up to 77 percent of Japan's total primary energy supply in 1973. As early as the late 1960s, before the oil shocks, Japanese society had already been faced with serious industry-related problems such as pollution, domestic market saturation, resource depletion, etc., all of which urged a shift toward a new structure such as information or knowledge-intensive industry. Japan's success both in transforming its industrial structure so it would be

TABLE II
Government Share of Total R&D Expenditure in Major Countries

		R&D (Billion Yen)	R&D for Defense (Billion Yen)	Government Share (%)	Government Share without R&D for Defense* (%)
Japan	(1982)	6,528	36	25.5	25.1
	(1983)	7,180	39	24.0	23.6
U.S.A.	(1982)	20,007	4,591	46.1	30.1
	(1983)	20,823	5,063	46.0	28.6
F.R.G.	(1983)	4,351	172	42.3	39.9
France	(1983)	2,643	564	57.8	46.3
U.K.	(1981)	2,694	754	49.8	30.3

$$* \% = \frac{(\text{Government R\&D} - \text{R\&D for Defense})}{(\text{R\&D} - \text{R\&D for Defense})} \times 100$$

Source: Kagaku-Gijyutus Yoran (A Handbook on Science and Technology), 1982–1983, Science and Technology Agency, Tokyo.

less reliant on energy-intensive industries and in developing energy conservation and oil-saving technology reduced the nation's industrial oil and energy input needs.

The second oil crisis in 1978 accelerated this tendency toward emphasizing less energy-intensive industries. Also, a huge trade deficit due to the oil bill led industries to concentrate on export gains. However, it is important to note that the oil crisis triggered a more fundamental change beyond the development of technology in the industrialized countries, especially in Japan. It triggered a transformation of the industrial structure to one based on information technology, thus promoting the beginning of the "information society."

In addition, during the period of the oil crises, a drastic change occurred in technological innovation at all levels of Japanese industry, including microelectronics, new materials, and software development. This technological thrust was different from previous types in various ways. Its most important feature was that industrial innovation was now driven not by focusing on a single component item of technology but by the so-called technological fusion or blending of several different functions known as mechatronics—i.e., a marriage between mechanical and electronic devices like robotics, computer-aided design and computer-aided manufacturing (CAD/CAM), and office automation. With this technological capability, Japan was able to undertake its industrial reorganization swiftly, thereby enhancing its international competitiveness.

In this new industrial pattern, one can see many new trends, including the regional diversification of high-technology sites (called the "techno-polis"); the introduction of the Flexible Manufacturing System, which allows for the production of many different products in small quantity; the entry of small companies, particularly in the software business; the revitalization of traditional industries with high technology; and the denationalization of public corporations. All of these trends rely, one way or another, on the application of high technologies.

The Current Situation. The race for survival among Japanese companies since the oil crises was quite intense and led most business leaders to the common view that unless they made constant R&D efforts, they would be eradicated from the market totally. This "survival consciousness" had serious international repercussions and resulted in an aggressive posture in international trade that eventually led to a backlash, known as "Japan bashing."

It is quite ironic that Japan rehabilitated its postwar economy as a result of being painfully conscious of its technology gap vis-à-vis the United States. Even more ironic is that after narrowing the gap, the severe economic setbacks caused by the two oil crises brought about a drastic industrial structural change in Japan. At the same time, the setbacks revived a new gap consciousness in Japan vis-à-vis the United States and somewhat vis-à-vis emerging competitors in the Asian newly industrializing nations (NICs), which are also involved with the United States in a high technology race for survival.

Despite its current trade deficit in some high-technology products, the United States is still predominant in "information technology," even in the civilian sector. Today, many Japanese industrialists are of the opinion that the United States has maintained its competitive edge in most of the crucial areas related to military technology, nuclear engineering, and aerospace such as microelectronics, computers, materials, and software. Therefore, if these technologies are applied to industrial production in the United States, Japan might not be able to survive as a major technological power. Such a new gap consciousness might generate a second round of U.S.-Japanese technological competition in the future.

This point is demonstrated in Table III, which shows Japan's competitive base in 80 key technologies as compared to the United States. The table was deduced from discussions at several committees organized by Japan's Ministry of International Trade and Industry, the Science and Technology Agency, and the Ministry of Welfare, as well as from the testimony of experts, public surveys, etc. The vertical axis indicates Japan's present level of technology in comparison with the United States. The horizontal axis, which compares Japan's R&D efforts with those of the United States, indicates how Japan is moving ahead progressively. Therefore, if, for example, there were a large cluster of cases at the upper-left corner of the table, this would indicate that Japan's present technological capabilities are more advanced than those of the United States but lacking in R&D intensiveness and therefore future potential. Most likely, this would mean that the technology had been imported and successfully assimilated. In actuality, Table III shows there is only one such case—the safety technology of the light water nuclear reactor. The table indicates two types of technology: the first includes those products with which Japan is equal or superior to the United States both now and in terms of future potential. The second type of technology represented is that in which Japan is inferior to the United States and includes products related to national security or military technology.

As a comparison, the gap between Japan and Europe is shown in Table IV. In contrast to the table for Japan and the United States, generally speaking, in this table, Japan is more advanced and more progressive than Europe, except for some technologies such as those for fast breeder nuclear reactors, lasers for medical purposes, resource exploration, and nuclear waste disposal. This indicates, again, that Japan is rather inferior

TABLE III
Japan's Technology Gap vis-à-vis the United States

JAPAN'S PRESENT STATUS	Least Progressive	Less Progressive	Equal	More Progressive	Most Progressive
Most Advanced					
More Advanced		1 technology: safety of lightwater nuclear reactor (LWR)	9 technologies such as optical fibers; industrial robots; fermentation; semi-conductors; videodiscs	4 technologies such as copiers; magnetic levitated trains	
Equal		5 technologies such as moving communication systems; machine tools; automatic interpreters	19 technologies such as large-scale computers; integrated circuits; optical communications; sensors	4 technologies: laser printers; artificial hearts; earthquake prediction; construction	
Less Advanced	3 technologies: CAD/CAM; medical lasers; seabed oil production	24 technologies such as space communication; microcomputers; nu-clear fusion; uranium enrichment; rockets; in-tragenic recombination	11 technologies such as coal liquefaction; extra high voltage power; control of crop production		
Least Advanced	1 technology: safety assessment of chemical materials	2 technologies: resource exploration; civil airplanes	1 technology: medical R&D		

JAPAN'S FUTURE POTENTIAL

Source: White Paper, Science and Technology Agency, Japan, 1985.

TABLE IV
Japan's Technology Gap vis-à-vis Europe

JAPAN'S PRESENT STATUS	Least Progressive	Less Progressive	Equal	More Progressive	Most Progressive
Most Advanced			2 technologies: optical fibers; videodiscs	2 technologies: optical elements; semiconductors	3 technologies: such as disaster prevention; copiers
More Advanced			8 technologies such as industrial robots; automatic interpreters; safety of LWRs	9 technologies such as sensors; optical fibers; fuel cell power generation	4 technologies: large-scale computers; earthquake prediction; artificial heart; LSI
Equal		6 technologies such as space communications; artificial satellites; seabed oil production	25 technologies such as CAD/CAM; rockets; nuclear fusion; intragenic recombination; fermentation; crop breeding	9 technologies such as laser printers; photo power generation; cattle breeding	3 technologies: medical R&D; housing construction; vermin prevention
Less Advanced	1 technology: medical lasers	6 technologies such as resource exploration; radioactive waste disposal; weather surveys	4 technologies such as civil airplanes; uranium enrichment		
Least Advanced	1 technology: fast breeder reactor				

JAPAN'S FUTURE POTENTIAL

Source: White Paper, Science and Technology Agency, Japan, 1985.

in the area of technologies that might relate to military or national projects such as space exploration.

As shown in these two tables, the present technology gap between Europe and Japan can be interpreted as a part of the technology gap between Europe and the United States. Japan's desire to catch up with the United States eventually resulted in its moving ahead of Europe in the civilian sector. However, one must pay attention to the fact that today the distinction between military and civilian technology is becoming obscure in highly advanced technology and that "dual-use"—namely, civilian and military application of the same item—is increasingly widening. This means that the technology gap in civilian technology can have a substantial impact on the capacity of military technology.

The International Landscape

As mentioned earlier, throughout the entire period of postwar Japanese history, the nation regarded the technology gap as a motivating force for catching up with the advanced industrialized countries. However, Japan's target was really limited to the United States, as the latter held an overwhelming lead in the high technology field. Europe, on the other hand, unlike during the Meiji period, was not necessarily regarded as a target. The statistics on the trade balance in major high-technology products clearly indicate the increased advantage of Japan vis-à-vis European countries in civilian technology, especially after the oil crises (see Charts I and II).

Chart I is the country share of trade in technology-intensive products (e.g., machinery, electronics, pharmaceuticals and chemicals) of Japan, West Germany, the United States, France and the United Kingdom. It is evident that Japan's share, having surpassed the 50 percent level, is increasing, while the share of other countries is either unchanged or even declining. However, if we add more items such as computers, semiconductors, aircraft, and video recorders, the ranking by country is different from Chart I. Chart II indicates U.S. preeminence, yet it is impressive that Japan is quickly catching up with the United States in the absolute value of technology trade. However, if we were to add nuclear-related technologies and space technologies, Japan's position would decline.[1]

The Euro-Japanese technology gap, on the other hand, has been widening to Japan's advantage, as the cases of video recorders, automobiles, consumer electronics, and many other products demonstrate. Europe seems to regard this situation as a serious threat that may shake the foundation of its trade and economy. The trade gap is rather complex and difficult to discuss from the perspective of the scientific and technological capability of Europe and Japan only. This is because competitiveness in trade reflects Japan's overall strength in the civilian sector in such areas as marketing, production technology, management skills, product development, etc. While Europe may be strong in fundamental research and in its capacity to develop new technology, if technology is treated as a total innovation cycle, it is evident that Europe is falling behind Japan. However, as mentioned earlier, it is generally accepted that Europe has an advantage over Japan in military technology and technology in-

[1]Though Charts I and II show the data up to 1981, we should note the recent dramatic change in the U.S. high-technology trade balance. As a report submitted to the Joint Economic Committee of the U.S. Congress in 1986 titled *The U.S. Trade Position in High Technology: 1980–1986* shows, since 1984 the U.S. high technology trade balance has declined sharply—$27 billion in 1981 and $4 billion in 1985. The report even forecasts that in 1986 the United States will face its first full-year high technology trade deficit judging from the first two-quarters of the year. The report maintains that the decline is due to the high value of the dollar, the expansion of U.S. domestic demand for high technology imports and the restriction of U.S. high technology exports for reasons of national security. On the other hand, Japan's technology balance ratio (not only high technology products but all technology exports over imports) reached nearly 1.0 in 1984 and has been increasing ever since. These changes simply suggest that even though the U.S. position in high technology is preeminent, a fuller account of the technology gap that provides more insight into what is happening within the infrastructures of these trade figures is needed. For example, many Japanese firms are investing heavily in the U.S. market, and many American firms are establishing joint-venture types of research establishments in Japan. Therefore, a simple look at the technology gap by means of trade figures does not sufficiently explain today's so-called "borderless" interaction of high technology.

CHART I
Country Share of Trade in Technology-Intensive Products

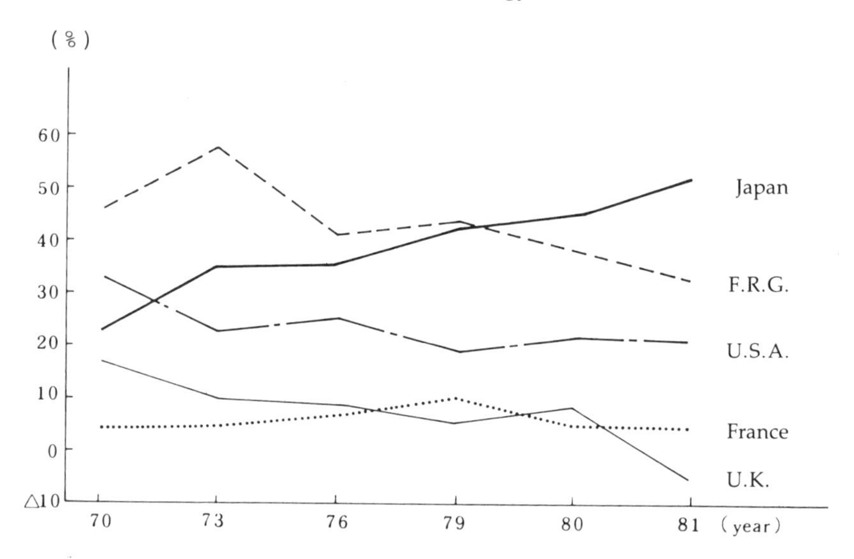

Source: Calculated from *Statistics of Foreign Trade,* Series B, Organization for Economic Cooperation and Development, Paris; *U.S. Exports,* Schedule E, U.S. Department of Commerce, Bureau of the Census, Washington, D.C.; and *U.S. Imports for Consumption and General Imports,* U.S. Department of Commerce, Bureau of the Census, Washington, D.C.

volving national projects such as aerospace, nuclear energy, and telecommunications.

Why was Japan able to assimilate high technology so successfully, at least compared to Western Europe? Can an answer be found in the managerial and social characteristics of Japanese firms? As it has been said, Japan is superior to Western Europe with regard to input factors of high technology such as R&D expenditures. However, as a comparison of Charts III and IV shows, even though R&D expenditure has some correlation with competitiveness in the international trade market, it is not the only determining factor. More important may be the nontechnical factors such as social attitudes and the response of workers to the adoption of new technologies. Evidence demonstrates that Japanese labor unions are more open to the in-

CHART II
Trade Balance of Major High-Technology Products

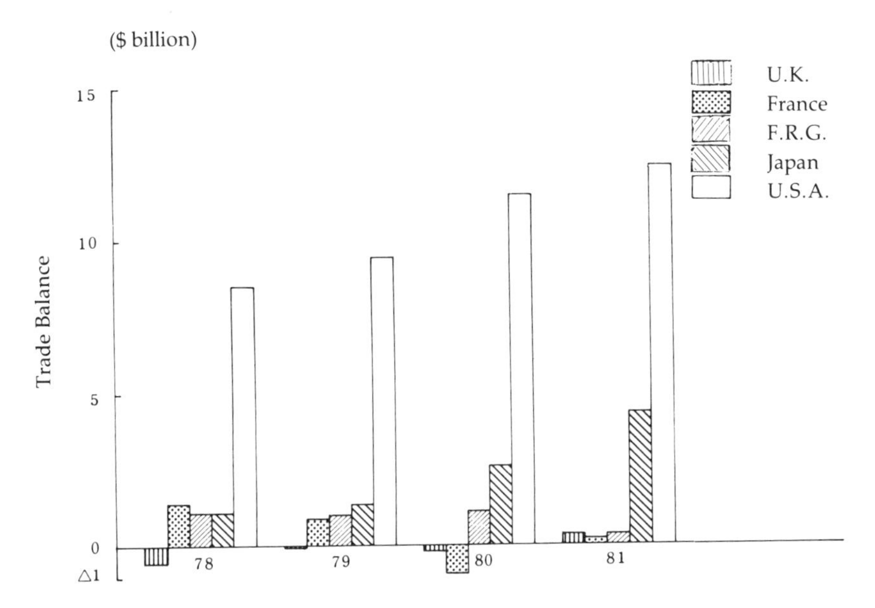

Source: Calculated from *Statistics of Foreign Trade,* Series B, Organization for Economic Cooperation and Development, Paris; *U.S. Exports,* Schedule E, U.S. Department of Commerce, Bureau of the Census, Washington, D.C.; and *U.S. Imports for Consumption and General Imports,* U.S. Department of Commerce, Bureau of the Census, Washington, D.C.

troduction of new technologies than European workers and that Japanese management has taken corresponding actions, including establishing a policy that guarantees lifetime employment for workers.

Lifetime employment was introduced in Japan shortly after World War II and was not necessarily part of the Japanese traditional system. During the period of economic reconstruction immediately after the war, there was a violent confrontation between labor and management. In an attempt to establish a better relationship with labor, industry adopted lifetime employment as a means of providing job security and profit sharing. Stability of the workforce was perceived to be a vital condition for recon-

CHART III
R&D Expenditure by Industrial Sector and its Share* in the Whole Industry

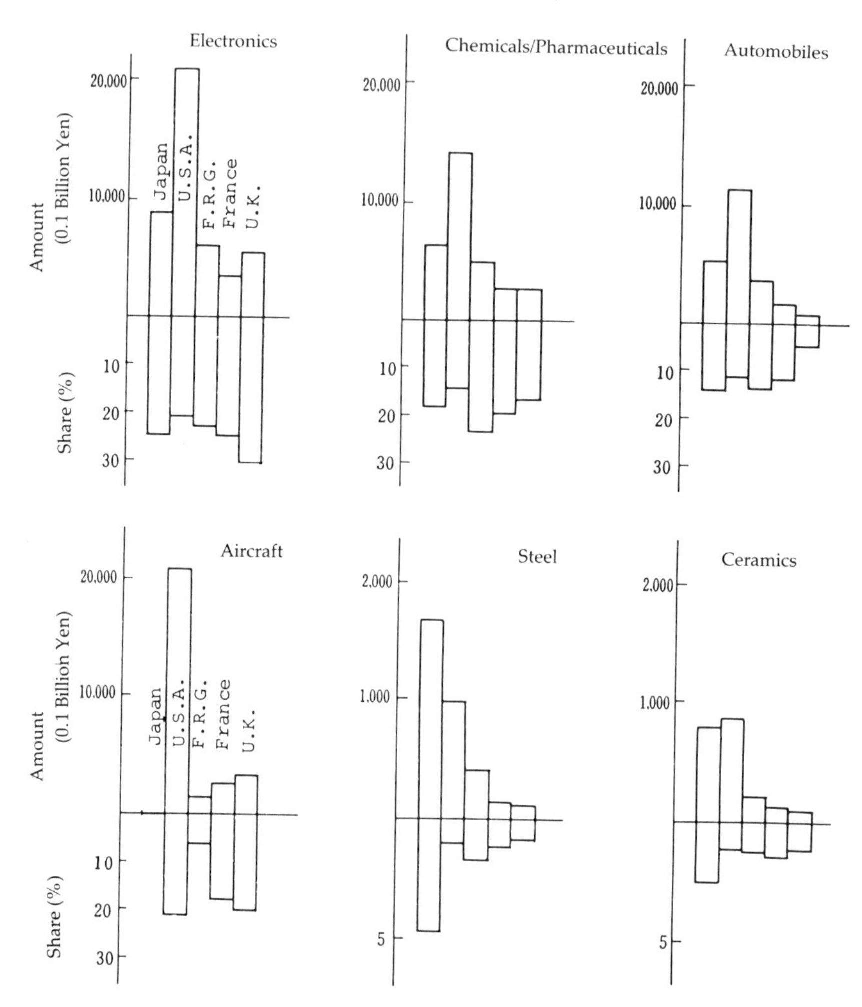

*Share = $\frac{\text{R\&D Expenditure by Industrial Sector}}{\text{R\&D Expenditure by Total Industry}}$

Source: White Paper on Science and Technology, Science and Technology Agency, Tokyo, 1985, p. 6.

CHART IV

Trade Balance Ratio (TBR)* of High-Technology Products

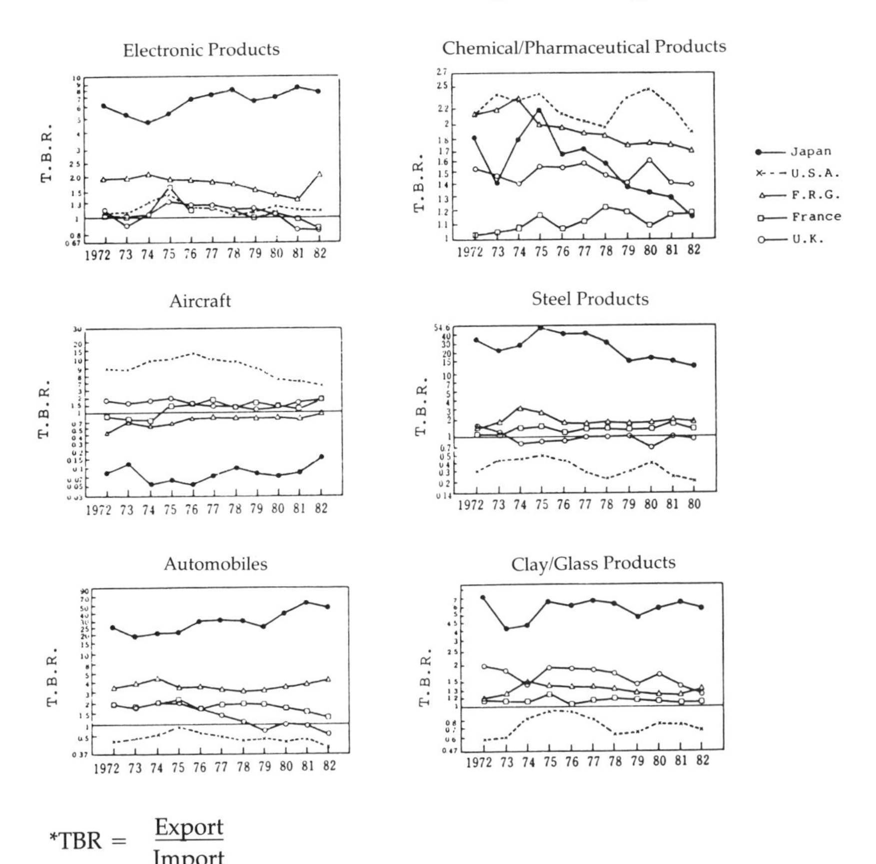

$$*TBR = \frac{\text{Export}}{\text{Import}}$$

Source: White Paper on Science and Technology, Science and Technology Agency, Tokyo, 1985, p. 7.

struction, especially crucial for the quick accumulation and assimilation of technology that was being introduced from abroad. Lifetime employment was designed to facilitate intra-company reorganization of the production system by securing worker stability at a time when foreign technology was being introduced in Japan. As a result, today Japanese labor unions are more flexible in introducing robots and other automated production schemes in contrast with European unions, which are traditionally antagonistic to new technology.

Some new developments pointing toward greater Euro-Japanese cooperation in the civilian sector have emerged recently. Joint efforts and technological cooperation are increasing between major firms; there has been some long-lasting cooperation. Continuing joint contracts between Philips and Matsushita and between Siemens and Fuji are two examples in the electronics field. Also, one can cite the long-standing and successful nuclear cooperation at the government level between Japan on the one hand and the United Kingdom, France, the Federal Republic of Germany, et al., on the other. However, in general, Euro-Japanese cooperation in technology among both private firms and governments is extremely weak compared to U.S.-Japanese cooperation. For example, even with the present stagnation in the U.S. nuclear industry, the General Electric-Toshiba/Hitachi and Westinghouse-Mitsubishi joint relationships dominate the Japanese field of light water nuclear reactors. Also, there is a much closer relationship in aerospace. All of this indicates how firm the technological alliance between Japan and the United States is as compared to that of Japan and Europe.

Given these facts, the technology gap between Japan and Europe must be analyzed keeping in mind that Japan has been entangled in the complex web of a technological alliance with the United States, predominantly in the civilian area but recently extending to military technology as well. This raises the question of how Europe will be able to deal with the emerging integration of economic relations within the Pacific region. In this regard, the rising competence in recent years of some Asian NICs should not be underestimated.

The emergence of the Asian NICs as competitors as well as partners to the United States and Japan has significant implications for the future of the North-North technology gap. For example, the ability of South Korea, Taiwan, Singapore and Hong Kong to catch up so rapidly in many high-technology areas is quite astonishing. Should their domestic firms develop an overall technological link with U.S. firms in areas such as software development, original equipment manufacturers

(OEM) supplies, and even assemblies, they could quite easily attain preeminence in a shorter period than anyone expects.

So far, for the most part, the Asian NICs have been playing the role of OEM suppliers to both the United States and Japan within the framework of U.S. technology and standards. However, soon they will have more freedom to supply other quality parts simply because they will have to expand their production and range of goods in order to benefit from economies of scale. This, in turn, means that it is conceivable that they will seek to expand beyond the trans-Pacific region to other parts of the world. At present, however, the Asian NICs are moving toward greater intra-Pacific integration. For example, General Motors and Chrysler are setting up an automobile production base in South Korea, and Texas Instruments and Matsushita Electric are making VSLI (very large-scale integrated circuit) investments in Taiwan.

Japan should recognize the significance of this integration and the role of the Asian NICs in it, instead of viewing the NICs as competitors attempting to take over Japan's share of the non-high-technology market, especially in the United States. The new industrial and technological integration in the Pacific area can have major significance for Euro-Japanese and Euro-American relationships, all the more if China joins the Asian NICs in the future. In any case, it is important to be aware of the dynamic changes in this region.

Although competition in the Pacific region among private firms will continue to be strong and complex, there is presently some change occurring in the Japanese perception of international technological relations. The accumulation of an unusually large trade surplus and the resulting international political pressure being put on Japan to change its industrial behavior in order to bring it in harmony with the rest of the world has caused serious concern in the Japanese government and in leading industrial circles. Consequently, industrial strategy for technology development has become one of the most crucial issues facing the Japanese government.

While the majority of experts believe that Japan is still technologically behind some major countries, especially the United States, in critical areas, many argue that Japan, as a major economic power, has to fulfill its responsibility in the international community by playing an active role in harmonizing world prosperity. The Japanese government insists that Japan's catch-up era is now over; that it should contribute as a leading industrialized nation to the accumulation of international common scientific and technological knowledge; and that it has to reverse its role from a technological borrower to a giver by transferring Japanese technology to other countries—to Western Europe and even to the United States.

Coincident with such government policy changes, Japanese industrialists, especially those involved in activities on a world scale, are assuming an internationalist view over the prevailing nationalist one. They assert that Japanese industries will be totally ousted from the international business community unless they seek joint prosperity with competing nations. They further argue that Japan's traditional "catch-up syndrome" should be discarded as it is too obsolete to apply to the current situation.

It is extremely difficult to forecast with confidence how Japan will approach the technology gap that exists vis-à-vis Europe in civilian areas. Yet, whatever course it takes, Japan must eventually consider how it and Europe can cooperate effectively in light of the complexity of the political, economic and technological relationships emerging in the trans-Pacific area. Vital factors will be how Japanese industrialists see the international technology gap on the whole and the role Japan will play as a key member of the world community.

Policy Implications

A characterization of the movement of technology across national borders should give some conceptual clarity to the policy implications of the technology gap. Science is fundamentally international and free to flow in its nature; technology, on the other hand, is proprietary and restrictive. However, with the increasing internationalization of economic activities, technology is "flowing" across borders extensively just like any typical economic good. Such flows of technology form "horizontal" links between countries, often conflicting with the political dimension in which the "vertical" concept of national boundaries and territoriality is still very much alive. In the latter case, technology is treated as a "pillar"—that is, as an instrument of the national power.[2] The current techno-political problems in the international domain often reflect the fundamental conflict between these two dimensions.

There are three policy implications that can immediately be deduced from the above discussion. The first concerns the role played by the *private sector*. The second involves the role of *government*. The third deals with the *political implications* of the technology gap.

In the *private sector*, it is encouraging to observe that the number of cases of Euro-Japanese joint ventures is increasing rapidly. The case of Siemens and Toshiba extends from the historical relationship between Siemens and the Fuji group. The case of Honda and British Leyland is an almost classic example of success. European and Japanese companies are energetically seeking to cooperate in the computer software field as well.

It has been noted already that some of the Japanese companies are still adhering to a catch-up, competitive philosophy and thus are hesitant to be open-minded about taking concerted action with international competitors. However, if in the future new products eventually require an international scale of standards on the one hand and compatibility to meet with versatile local needs on the other (as in the case of computer software or VHS video cassettes), companies will not be able to confine their activities to the domestic market only. Rather, they will have to meet international standards that will tend more and more to reflect the social and cultural practices of local regions. Otherwise, they will be gradually ousted from the international marketplace. Therefore, international cooperation with local firms will become more important, even for market strategy. That is to say, cooperation among firms beyond national boundaries will serve as insurance, as a hedge for survival.

At the same time, if European countries are seriously considering the rehabilitation of their industrial competence vis-à-vis

---

[2] I am especially indebted to Ellen Frost for her insightful comments on this subject.

Japan, one thing they must consider is how to better deal with the intra-Pacific technological alliance that includes the Asian NICs. It is foreseeable that the development of industries will be stimulated by technological cooperation, eventually leading to company-to-company cooperation in the form of joint ventures.

As to the role to be played by *government*, the major question is how it can respond to spontaneous actions taken by the private sector. What government can and should do is to lay out the infrastructure to assist the private sector. It is also the role of government to encourage a shift in social attitudes to the point where the public wants to close the technology gap. Another task for government is to provide support for a qualified workforce. Finally, government should recognize that the gap, if left to market forces, will eventually be dissipated as technological cooperation grows among firms in the private sector that share commercial interests, as recent trends reported by the Japanese External Trade Organization (JETRO) have indicated.

The most important and yet difficult aspects of the technology gap between Europe and Japan lie in the *political* realm. It has been stated repeatedly that Euro-Japanese cooperation needs to take into account the trans-Pacific alliance between Japan and the United States, even though cooperation is civilian in nature. The current trade politics between Japan and the United States is centered on Japan's huge trade surpluses. But curiously enough, both governments are more or less in sync with each other despite the many unresolved problems between the two countries. Their decision to impose auto export quotas for more than five years on Japanese cars entering the United States is one good example.

In the private industrial sectors, U.S.-Japanese cooperation is growing much faster than most people believe. Recently, the JETRO reported a rapid increase in U.S.-Japanese industrial cooperation, such as joint ventures, joint development and technological exchange. In 1985, 652 cases of such cooperation were reported compared to 390 in 1984, a 67.2 percent increase. The greatest amount of cooperation came in the high-technology sector where 242 cases were reported involving such products as computers, semiconductors and integrated circuits. Industrial cooperation between the United States and Japan and the Asian

NICs is increasing as well. The fact is that because Japan and the United States are strongly bound to each other as political, economic and technological allies, it is difficult to formulate a cooperative scheme between Japan and Europe.

Another crucial political implication of the technology gap between Europe and Japan is its impact on the national security of the European countries, especially with the increasing "dual use" of technology. Whether or not the weakening of some European industries through commercial competition can threaten the basis of Europe's military technology is a fair question. One major difficulty on the Japanese side in this regard is that present policy forbids any export of or cooperation in military technology, except with the United States. Therefore, up to now, there has been no possibility of cooperation in military technology between Europe and Japan.

In many ways, the present situation in international technology urges us to reassess our historical trilateral relationships taking the new wave of technological innovation into consideration. If Europeans see that their weakened industrial bases could endanger their security position vis-à-vis the Eastern bloc, Japan and Europe should seriously review their cooperative relationship keeping this political framework in mind. Partly due to the annual summit conferences and partly because of Japan's full-fledged R&D in such national projects as aerospace and nuclear energy, the government-to-government cooperation has gradually been expanded beyond the U.S.-Japan bilateral pattern.

There are many unresolved problems in the Euro-Japanese technology relationship, such as how to build up common understanding with respect to the technology gap, or whether Japan can promote human contact with Europe by shifting the direction of its educational exchanges away from the United States and more toward Europe. Despite the many difficult issues, Japan is certain that it should construct a new policy paradigm to encompass new aspects of international technological cooperation. Foremost in this is Japan's desire to continue to reduce its once heavy dependence on U.S. technology.

Japan is now a target of international criticism. It is under pressure to develop a completely new scheme of international technological exchange that deals simultaneously with many problems such as: enhancement of basic research; adoption of an open door policy; introduction of tax reforms to eradicate Japan's preferential treatments associated with technology transfer; participation of foreign companies in Japanese government-sponsored research and projects; release of technical information to other countries; and even regulation of excessive competition. It is now clear that a piecemeal approach to solving each of the above problems is futile. What is needed is strong leadership and positive initiative to construct a new Japanese policy for international technological exchange.

The first step toward such a policy will be to develop a Japanese "vision" of international relations and identify where Japan is located in it. In order to do this two core issues are fundamental and must be addressed:

1) What will be the future U.S.-Japan relationship in both industrial and military technologies? How will this extend to relationships with NICs in the trans-Pacific context? Should Japan stay as a predominantly civilian power or step more actively into the military realm?

2) What will be the Euro-Japanese relationship as each region tries to realize harmonized industrial development in the context of economic prosperity and national security? Is there a special role to be played by a more integrated Europe?

Furthermore, the relationship between the industrialized countries and the Third World, and the role Japan can play is an urgent issue for future world political and economic stability, though this is not directly related to the theme of this essay. Until recently, technology has been used as a major vehicle for prosperity in particular countries, but nowadays it is becoming a vehicle for prosperity and improved human welfare on a global scale. It is important that we create international understanding and a common political basis among the "trilateral nations" to provide for an acceleration of international technology exchange.

I am especially indebted to Taizo Yakushiji for his assistance in preparing this manuscript.